EXTINCTIONS

EXTINCTIONS

From Dinosaurs to You

Charles Frankel

The University of Chicago Press
Chicago and London

The University of Chicago Press, Chicago 60637
The University of Chicago Press, Ltd., London
© 2024 by The University of Chicago
Published 2024
Printed in the United States of America

33 32 31 30 29 28 27 26 25 24   1 2 3 4 5

ISBN-13: 978-0-226-74101-7 (cloth)
ISBN-13: 978-0-226-74115-4 (e-book)
DOI: https://doi.org/10.7208/chicago/9780226741154.001.0001

Library of Congress Cataloging-in-Publication Data

Names: Frankel, Charles, author.
Title: Extinctions : from dinosaurs to you / Charles Frankel.
Description: Chicago : The University of Chicago Press, 2024. | Includes index.
Identifiers: LCCN 2023037667 | ISBN 9780226741017 (cloth) | ISBN 9780226741154 (ebook)
Subjects: LCSH: Extinction (Biology) | Mass extinctions.
Classification: LCC QH78 .F73 2024 | DDC 576.8/4 — dc23/eng/20230921
LC record available at https://lccn.loc.gov/2023037667

♾ This paper meets the requirements of ANSI/NISO Z39.48-1992 (Permanence of Paper).

Contents

Introduction

Our civilization is confronted today with a major breakdown of the biosphere. Many species are quivering on the edge of extinction, including elephants in Africa, tigers in Asia, lemurs in Madagascar, antelopes in Kazakhstan, and sea otters along the North Pacific coast.

These emblematic species make up only the tip of the iceberg. Across the world, birds are declining, amphibians are crashing, and insect species are vanishing before we can even catalog them—to the point where observers now describe the ongoing crisis as marking the beginning of the sixth great mass extinction on Earth, threatening to match the "Big Five" that rocked the planet's biosphere in the past. The last crisis of this kind, 66 million years ago, took out the dinosaurs and three-quarters of all species on land and in the oceans in the aftermath of an asteroid collision with our planet.

If we do not rectify our current situation, the ongoing collapse of the ecosystem might well measure up to this end-Cretaceous mass extinction, both in terms of the number of casualties and the abruptness of the crisis. This book aims to place the present situation in a global perspective. After reviewing the lethal mechanisms of past extinctions, it examines which ones are at play today, how many species have gone extinct under our watch, how many are threatened and critically endangered, and what solutions we can bring to the problem. Much in our assessment of the situation and our prediction of the outcome relies on the understanding of how a crisis builds up over time, and how feedback loops can worsen the situation.

Fortunately, we do not start with a blank page. Much can be learned from past mass extinctions in order to better assess the ongoing crisis. Hence, the opening chapter reviews a number of past mass extinctions on record, and although their settings and conditions were different from those today, many extinction mechanisms are relevant with respect to the present crisis.

In chapter 2, I chose to focus on the fifth mass extinction, the end-Cretaceous crisis, for several reasons. Since it is the most recent extinction, at 66 million years ago, it holds the best fossil record of how the biosphere collapsed, and it brings into play several mechanisms that concern us today—a breakdown of the food chain, global warming, and ocean acidification. It also ushers in our contemporary Cenozoic era ("modern life" in Greek), which frames the rise of birds, mammals, and ultimately the human species: an era that the current crisis might well bring to a close.

There is another reason to emphasize the end-Cretaceous mass extinction that is sociological in nature. At first, many scientists downplayed its scope and its brutality, and argued that dinosaurs and other species had died out progressively over millions of years, rather than abruptly, denying the notion that the biosphere could collapse catastrophically. Today, there are scientists and opinion makers who minimize the impact of humanity on the environment, deny the amplitude of climate change, and downplay the number of species threatened with extinction.

Chapter 3 describes the recovery of the biosphere after the end-Cretaceous collapse. Starting over from a cut-down and reshuffled deck of cards—a reduced cast of surviving animals and plants—evolution played out a new game of adaptation and survival, thrusting birds and mammals into the limelight and onto a course that experienced its own share of minor crises. These included an abrupt warming spike 55 million years ago that altered the course of evolution and a subsequent cooling trend that began 35 million years ago with the glaciation of Antarctica. Finally, over the last 10 million years, temperature and rainfall variations affected the East African Rift and might have played a decisive role in steering the evolution of the human genus toward a bipedal posture and the toolmaking revolution. This chapter is somewhat technical and overly detailed in places, so I would suggest skimming through the parts that appear too academic—except students of the earth sciences, of course!

Chapter 4 picks up the story when our species rose to prominence and spread across all continents, from 50,000 to 15,000 thousand years ago: a crucial time when most large mammals that roamed the planet—mammoths, woolly rhinoceroses, and giant sloths, to name but a few—went extinct. The collapse of the megafauna was long attributed to the warming trend that brought the last ice age to a close, shrinking habitat and food, but the coincidence in time with the expansion of human hunters also makes a strong case for the direct responsibility of *Homo sapiens*. If overhunting by our species is to blame, the collapse of the megafauna marks the starting point in the geological record when we manifest ourselves as a disrupting agent, altering the course of evolution and leaving an indelible mark in Earth's history. Although the megafaunal collapse is spread over several thousand years, it is very brief by geological standards. It can be argued that it marks the beginning of the Anthropocene: the new geological epoch that scholars define as showing the imprint of human civilization on the planet.

After a brief respite, during which its influence on the biosphere is less apparent, our species triggered a second spike in the extinction record. Chapter 5 describes the dire consequences of the great voyages of discovery, as Westerners sailed across the Pacific and Indian Oceans: across the islands, vulnerable species were decimated. Such massacres were portrayed at first as colorful anecdotes in which clumsy and inept species "deserved" to go extinct, as with the dodo bird of Mauritius Island. We know today that the fate of these species was sealed not by some inherent weakness or poor adaptation to the natural world but by the inescapable pressure of irresponsible mariners: the landing of hungry explorers, accompanied by hunting dogs and stowaway opportunistic rats, might have exterminated close to one thousand bird species (10 percent of Earth's avian diversity) in less than two centuries.

Chapter 6 brings us to the core of the present crisis: after the raid on ocean islands, entire continents suffered the combined assault of overhunting, poaching, deforestation, and agriculture, as well as of invasive species that were carried around the world. In this pivotal chapter, I attempt to estimate the number of endangered species and extrapolate how long it might take us to reach the casualty figures of a true mass extinction.

Linear extrapolations of current trends, however, might underestimate future damages to the ecosystem. As discussed in chapter 7, the ongoing crisis is compounded by a series of aggravating factors that might steepen the killing curve. Climate change is one such factor that affects both land and sea. On land, animals are already on the move, struggling to stay ahead of the heat wave that sweeps across the continents. The oceans are also feeling the heat, losing their coral reefs and, by absorbing a fraction of our surplus carbon dioxide, experiencing a damaging rise of acidity.

Is there a way to reverse, or at least to mitigate, the ongoing collapse of the biosphere? Chapter 8 hints that the situation is far from desperate—except for the species that are already extinct—but strong action must be taken now, and at all levels of society. Politicians are not the only decision-makers who can tackle the situation; we all have a role to play in the style of living we embrace, from our consumption of goods and energy to our feeding habits. Rising to the challenge, concerned citizens, organizations, and governments are implementing clever solutions worldwide.

Should we overcome our many problems and contradictions, and successfully fight off today's looming mass extinction, there is no guarantee that our civilization and species will dodge that bullet forever. A speculative closing chapter addresses the destiny of humanity. How long can we expect to last? What other doomsday scenarios might we encounter and overcome in order to survive several more centuries or millennia, or if we are very wise and lucky, hundreds of thousands of years? Might we even leave our planetary cradle—whatever shape we leave it in—and spread across the galaxy?

When I began to do research for this book, I had no clear idea of the numbers and trends I would end up finding. Were we really going down the path of a major mass extinction? Without pretending to be objective, I was open to any conclusion, and I hope the reader will enjoy this book as an investigation, rather than a demonstration thought out in advance. It also profits from two phases of research and writing: I penned a first version for my French publisher, Le Seuil (Paris), in 2016, under the title *Extinctions: Du dinosaure à l'homme*. Writing a revised, updated version for the University of Chicago Press was an exciting challenge: it gave me extra time to broaden my research—there are so many relevant articles to absorb—and to update the numbers, refine trends, and improve predictions.

The endnotes for each chapter give extra details on a topic or offer leads to academic research articles and official reports—the note section serving as a succinct bibliography. I deeply apologize for not citing all the great work that I consulted. Likewise, I introduced by name into the writing only a limited cast of scientists, when so many deserve to be portrayed; I occasionally mention their nationality, to give a sense of the multinational dimension of the scientific community. Charts and illustrations are mostly derived or adapted from research articles, and many photographs were kindly provided by researchers and protectors of the environment.

I would like to express my gratitude to my editors at the University of Chicago Press: Susan Bielstein for her support and patience as we began this project together, and Joseph Calamia, assisted by Matt Lang, for bringing it to fruition. Many thanks to Carrie Love for copyediting my manuscript, polishing my English, and suggesting clarifications and improvements, making this a better book, to Stephen Twilley for overseeing the production stages, and to book designer Ryan Li and all the staff involved in its layout. Happy reading, if there is such a thing as happiness when one dives into extinctions! May we all improve our stewardship of planet Earth.

1

Welcome to the Anthropocene

There is a consensus today, among concerned scientists and observers of the natural world, that the Earth's biosphere is in trouble. The complex network of animals and plants, held together by an intricate web of relationships, threatens to collapse under the mounting pressure of human activities: the explosive growth of our population—threefold in less than a century—accompanied by an expansion of agriculture and deforestation, industry, overhunting and overfishing, global warming, and seawater acidification.

Modern civilization is leaving such a distinct mark on the planet, be it chemically or in the fossil record, that if an extraterrestrial civilization were to visit Earth millions of years from now, it would have no trouble identifying the precise level in the sedimentary record where humankind started tampering with the environment. Sediments that pile up on the bottom of seas and lakes indeed record a number of parameters—not only mud type and fossils but also intricate chemical details, including the acidity and temperature of water at the time.

Arguably, in the sedimentary record, the impact of humankind is recognizable as early as 15,000 to 10,000 years ago, if we take responsibility for the extermination of the megafauna—mammoths, giant sloths, and the like—that took place at the close of the last ice age. Such a brief interval of time amounts to little more than a few centimeters (an inch or two) of sediment deposited on the deep ocean floor. Hence, the damage that the human species has inflicted on the biosphere over the past 10,000 years will leave a boundary as narrow and sharp as the couple of centimeters (1 in.) of clay of the end-Cretaceous mass extinction 66 million years ago, when three-quarters of all species on Earth disappeared. Our goal, of course, is to halt the tally so that the couple of centimeters of sediment that mark our tampering with the biosphere never come close to matching the devastation of the end-Cretaceous crisis.

Close up, the sediment layer that marks the onset of the Anthropocene—stretched out to more than a few centimeters if we look at a "high fidelity" accumulation of mud in a lake or a shallow sea—points to several major events. The first is a sharp temperature drop, known to climatologists as the Younger Dryas event, around 12,800 years ago. There are many theories about what triggered the abrupt cooling; one in particular blames it on the disappearance of the megafauna, animals that contributed through their metabolism a large amount of methane—a strong greenhouse gas—to the environment. The demise of the large beasts, which caused a sudden drop in atmospheric methane, might have interrupted the warming trend that was pulling the Earth out of the last ice age. This interesting hypothesis, suggesting early human tampering with the climate, will be further explored in chapter 4.

A second event that shows up in our Anthropocene boundary layer is the onset of agriculture: a pollen spike of cultivated species starting 11,000 years ago, followed by small methane, carbon dioxide, and temperature swings that reflect phases of deforestation, agriculture, rice paddies, and occasional reforestation. The top of the Anthropocene layer also shows a steady temperature rise, reflected in its isotopic chemistry, accompanied by a simultaneous surge of carbonaceous particles that leaves no doubt as to its origin: the massive burning of fossil fuel.

Another imprint of humankind, superimposed on the carbon spike, is a fine sprinkle of radioactive decay elements—principally strontium, cesium, and americium—caused by atomic bomb testing in the atmosphere from the late 1940s up to 1965. The spike is so sharp and unnatural that many scientists propose it to be the official boundary of the Anthropocene, since it is easy to spot in sediments worldwide, although our influence on the environment probably began thousands of years earlier.

Just above the radioactive spike, a second unmistakable imprint of human civilization is the surge of microplastics. Discharged into the environment today, these plastics are bound to settle on the ocean bottom over time and will be incorporated into the sediment column. Other elements and clues will end up in the sedimentary record over the years to come. How the Anthropocene boundary ultimately terminates, and what new world emerges above it, is anyone's guess. In the worst-case scenario, the boundary could be capped by a second radioactive spike, as civilization brings itself to a close through nuclear warfare. In the best of worlds, the boundary might transition to a new geological epoch, where the environment stabilizes and biodiversity recovers, as human civilization adopts a more balanced approach to its home planet.

Extinction as a Notion

The big question is how many plant and animal species will have gone extinct by the time we readjust the balance. Will the number of exterminated animals and plants compare to the toll of other great mass extinctions of the past?

As we ponder this question, it is worth noting that the very notion of mass extinctions is a fairly recent concept. The idea that species are mortal in the first place was long ignored in Western culture, be it in science or religion. According to Judeo-Christian belief, the creation of the world had to be perfect, otherwise it would mean that the Creator had failed in his undertaking. Even in his greatest fury, as recounted in Genesis, all species were spared during the Great Flood, although the prerequisite to board Noah's ark was to be a "pure" species, which leaves some room for interpretation.

Starting in the eighteenth century, the ruling dogma was chal-

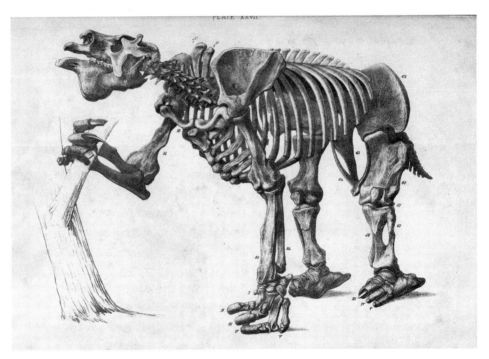

Scientists long believed that animal species known only from their fossilized bones—such as the giant sloth of the Americas (shown here)—were not extinct but simply in hiding, awaiting discovery. From Richard Owen, *Memoir on the Megatherium* (London: Williams & Norgate, 1861), through the Biodiversity Heritage Library.

lenged by the discovery of spectacular bone beds—those of mammoths and other giants of the Ice Age—that did not correspond to any living animals. If these were truly extinct species, then the Creator had let them perish, recognizing that they were failures. A number of scholars hung on to the hope that these unknown species were not extinct but simply missing, and that living representatives would ultimately be discovered in remote locations on Earth, where they had sought refuge.

Hence, European hunters were on the lookout for the giant elk *Megaloceros*, with antlers 3 meters wide (10 ft.), fossils of which had been discovered in peat bogs from Ireland to China. Likewise, American president Thomas Jefferson (1743–1826)—an avid fossil collector—instructed Lewis and Clark, explorers of the West, to bring back a living specimen of a giant sloth, or if they were so lucky, a couple of live mammoths as well.

As time went by, hope dwindled, and the paradigm of a fixed and perfect creation became increasingly challenged as the number of fossil species kept growing: extinctions seemed to be a real trend in the natural world. Moreover, as they sailed across the Pacific and Indian Oceans, explorers were confronted with the process firsthand: they enacted and witnessed the elimination of defenseless bird species as they overhunted every island they reached. Hence came the realization that the Earth's biosphere was not a permanent cast of animals and plants, but rather a dynamic system in which new species progressively developed, out of genetic reshuffling among their members, while others vanished.

How Many Species Are on Earth?

One would imagine that with the scientific and statistical tools at our disposal, we could easily assess the number of species that are alive today. This is hardly the case. Even with respect to mammals and birds—animals that are difficult to miss—the count is not over yet.

Consider the mammal class: only 3,000 species were known in the early 1900s. The count has doubled a century later (breaking the 6,000 mark in 2019), and about two dozen new species are added to the list each year. The census is also far from complete when we turn to plants: 350,000 species have been described, but researchers believe at least 100,000 are still awaiting discovery. Other categories are much less complete: although 1 million insect species are currently registered, at least another 5 million remain to be discovered, according to experts.

Even more challenging is the census of marine species, because of the cost and complexity of research at sea, be it in coral reefs close to land or in deep waters across abyssal plains and mid-ocean ridges. Over 230,000 marine species have been described so far—from tiny plankton to large fish and sea mammals—and another 70,000 collected species are on the waiting list, aboard research vessels and in laboratories around the world, waiting to be analyzed and cataloged. But these figures make up only the tip of the iceberg: the true number of marine species is believed to fall somewhere between 1 million and 2 million. As for fungi—mushrooms and molds—they are among the least described group of living beings: only 150,000 species are cataloged, while their number is estimated to range between 2 million and 4 million, meaning that 95 percent have eluded scientists so far.

And what about bacteria? Because they are microscopic and unicellular, they tend to be overlooked in the appraisal of Earth's biodiversity. This is inconsiderate, since bacteria likely represent over one-half of the planet's biomass (the mass of all living beings) and were its first inhabitants, giving rise to all other multicellular plants, molds, and animals. Despite their importance, little more than 10,000 bacteria species have been described: only a drop in the bucket. A bottle of seawater is estimated to contain close to 20,000 distinct species, and each vertebrate animal lodges several hundred to several thousand specialized bacteria involved in symbiotic or parasitic relationships with their host. Research has naturally focused only on human bacteria and on those of our most important domestic animals, hence the small tally so far, but microbiologists estimate the total number of bacteria species to be about 1 million.[1]

Viruses are usually ignored in the count since they do not qualify as true living beings: they are simple strands of nucleic acids (DNA or RNA) encapsulated in a protein shell, incapable of growing, reproducing, or even functioning outside of a host cell, which provides them with their most basic vital functions. However, one would guess that the number of viral "species" is about a million as well.

In conclusion, then, and ignoring bacteria and viruses, approximately 2 million species of animals, plants, and fungi are described today, and biologists add 15,000 to 20,000 new species to the list each year.

How long will it take to complete the great catalog of life on Earth? If naturalists do not receive additional help and keep cataloging at the present rate, completing the job might take three to

Species, Families, and Classes

Classifying species is a complex task that involves giving each species a name, or *taxon*, and figuring out its relationship to others—its ascendants and descendants. This is the science of *phylogeny*. To make things simple, one can represent biodiversity as a family tree, with branches splitting over time into multiple offshoots.

The tree of life diverges initially into three major directions: bacteria (cells without a nucleus), archaea (similar to bacteria, but with distinct DNA and RNA, and a different cell membrane), and, finally, eukaryotes (cells containing a membrane-protected nucleus). Eukaryotes interest us most, since we belong to that group. They split into half a dozen large branches, three of which are particularly familiar to us: plants, animals, and fungi. In turn, each of these branches split over time into different boughs. Hence, animals diversified into cnidarians (such as jellyfish), chordates (vertebrates and the like), arthropods (insects and crustaceans), and mollusks (such as snails and squids), to name but a few.

Each one of these boughs then radiates into yet finer branches, named *classes*. Vertebrates, for example, include several fish classes, as well as amphibians, reptiles, birds, and mammals. Classes then break up into various *orders*; there are close to twenty in the mammal class, such as chiropters (bats), cetaceans (whales, dolphins, and porpoises), carnivores (wolves, cats, and the like), marsupials, and primates.

An order, in turn, splits into *families*. The primate order includes tarsiers, ring-tailed and woolly lemurs, capuchin and squirrel monkeys, Old World monkeys, and great apes (Hominidae family).

Reaching the finest twigs of the phylogenic tree, families then split off into *genera*. Great apes comprise four genera: orangutans, gorillas, chimpanzees and bonobos, and modern humans and their close relatives (*Homo* genus).

Finally, genera radiate into individual, distinct species: only one is left in the *Homo* genus—*Homo sapiens*—while gorillas include both a West African species and an East African one.

four centuries, based on the estimate of unknown species: a figure believed to fall somewhere between 7.5 million and 10 million. If this figure is accurate, only one-quarter of all living species have been identified so far.[2]

Looking Back in Time

The present tally is only a snapshot of what turns out to be a very dynamic turnover. According to estimates, the several million

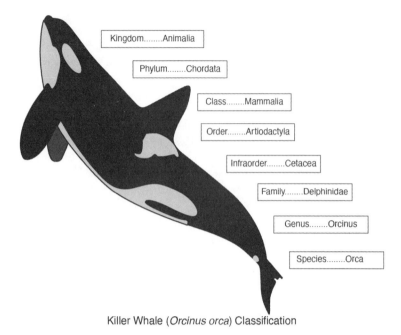

Killer Whale (*Orcinus orca*) Classification

Living species are classified into groups and subgroups, according to similarities in appearance and genetic analysis. The killer whale, for example, belongs to the oceanic dolphin family (Delphinidae), which itself belongs to the Cetacea infraorder of aquatic mammals.

species alive today represent only one-tenth of 1 percent (0.1 percent) of the total number of species that once roamed the planet. In other words, for every species alive today, one thousand have gone extinct along the way.

A species goes extinct if its members die off faster than they can reproduce, because of some trauma or lack of adaptation to a changing environment. In many cases, offshoots of a species will genetically mutate and evolve into new species before the mother stock goes extinct, keeping the phylogenic tree of life healthy and diverse.

In the same way that one can estimate the average life expectancy of an individual—several decades for a human being, several centuries for a sequoia tree—one can estimate the average lifespan of a species as a whole by looking at the fossil record. Dating the layer where a specific seashell, plant spore, or vertebrate skeleton first appears, and dating the last layer in which it is found, yields the species' lifespan: 4 million to 5 million years on average. The turnover is faster in some groups than in others: among mammals (our own class), the average lifespan is shorter, closer to one million years. Since the human species first appeared 300,000 years ago, there is some hope that our tenure on Earth is far from over, but averages are only a statistical tool and by no means a guarantee.

Besides this "background" turnover that constantly updates the natural world, species can also disappear together in droves, during particular crises that affect large sections of the biosphere. One of the first scientists to recognize these sharp inflections in the course of evolution was French paleontologist Georges Cuvier (1769–1832), who cataloged the occurrence and disappearance of fossil species in the sedimentary strata surrounding Paris. The scientist discovered that lake-bed sediments containing bones of mammal species abruptly switched to marine sediments containing fossil shellfish. To explain such upheavals, which he named "global revolutions," Cuvier imagined worldwide earthquakes and floods. After each major crisis, the Earth would repopulate, thanks to surviving species that had found shelter somewhere on the planet, or else through some new episode of divine creation.

Cuvier's concept that many species could vanish simultaneously—which became known as *catastrophism*—was criticized and challenged by those geologists who emphasized that the Earth could only change through slow, incremental processes, which they witnessed in the field. They mocked catastrophism as being overly spectacular, with biblical overtones, and banished it from respectable scientific circles.

One English geologist, however, came to the rescue of Georges Cuvier. Logging all fossil species known to him, John Phillips (1800–1874) established a geological timescale in which he recognized several distinct breaks in the succession of lifeforms. These breaks were so prominent that he divided the history of the Earth into three major eras: Paleozoic (the era of ancient life, in Greek), Mesozoic (the era of intermediate life), and Cenozoic (the era of recent life).

The first break, between Paleozoic and Mesozoic, is spectacular, with respect to the number of species that vanished: probably more than 90 percent. The break occurs, if we zoom to the finer scale of geological periods—each era is divided into periods— between the Permian (last period of the Paleozoic) and the Triassic (first period of the Mesozoic). It is therefore known as the end-Permian, or Permian-Triassic, mass extinction.

The second major break, between the Mesozoic and Cenozoic eras, is almost as profound. Since it occurs between the Cretaceous (last period of the Mesozoic) and the Paleogene (first period of the Cenozoic), it is known as the end-Cretaceous, or Cretaceous-Paleogene, mass extinction. This is when the dinosaur group went extinct.

John Phillips was ahead of the game. A century rolled by before mass extinctions were taken seriously. It wasn't until the 1950s that German paleontologist Otto Schindewolf (1896–1971) reemphasized the exceptional amplitude of the end-Permian mass extinction and attempted to explain it with just as exceptional a theory. He suggested that the explosion of a nearby star— a supernova—had showered the Earth and its biosphere with lethal cosmic rays.

Even if Schindewolf's supernova theory was never proven in the long run, it had the merit to shake up the scientific community and make geologists aware that the planet might be at risk, and that mass extinctions were worth examining in greater detail. Indeed, it wasn't long until American geologist Norman Newell (1909–2005) reopened the mass extinction files, combed over the fossil record, and in a seminal text published in 1967, highlighted five, rather than two, major breaks in the evolution of life on Earth, each characterized by a severe drop in the number of species across the boundaries.[3]

Newell identified the first break as being at the end of the Ordovician period, 445 million years ago, and the second one at the end of the Devonian, 375 million years ago. He confirmed the third one as the landmark end-Permian mass extinction, 250 million years ago; identified a fourth one at the end of the Triassic period, 200 million years ago; and endorsed the dinosaur-killing end-Cretaceous mass extinction, 65 million years ago (all approximate ages).

American paleontologist Jack Sepkoski (1948–1999) screened the fossil record even further. In the late 1970s, he pinpointed over a dozen smaller crises through a careful statistical analysis of the data, although many barely stand out over the background turnover of species and others appear to be only regional in nature.[4] The five largest mass extinctions, on the other hand, nicknamed the "Big Five," stand significantly above background rates around the world: across each boundary, over 50 percent of tallied species are shown to disappear.

The realization that the Earth's biosphere was rocked by such major upheavals found particular resonance among scientific circles in the 1970s, because it coincided with the burgeoning consciousness that human civilization, through overhunting, deforestation, and pollution, was sparking a new crisis of perhaps comparable proportions. In the decades that followed, emerged the concept of a human-driven, sixth great mass extinction.

As the figures in this book will show, we are still a long way from matching the previous crises, since the proportion of species extinguished by human civilization is currently estimated to be 2–3 percent, a far cry from the 50–75 percent needed to qualify as a great mass extinction. The pace of extinctions, however, is quickening, and we are now in the process of acknowledging the magnitude and urgency of the problem.

Looking back at the geological record is a good thing. Scientists have the opportunity to study what went wrong in the past, in particular to figure out how feedback loops came into play. They can study what types of species were most affected by each crisis and how the biosphere managed to mend its wounds and bounce back in the long run. Here is, in a nutshell, the sequence of events, main victims, and possible causes of the first great mass extinctions on Earth.

Before we turn our attention to the "Big Five," it should be said that other crises of the biosphere occurred before the first big ones we are about to discuss, but they took place at remote times in the past, when lifeforms had not yet developed the hard shells and bones necessary for their fossils to be preserved well enough for analysis. One such major crisis was the Great Oxidation Event, also called the Oxygen Catastrophe, that was triggered, approximately 2.4 billion years ago, by an unprecedented amount of oxygen being released into the ocean and atmosphere by a new brand of marine bacteria that had developed photosynthesis as an energy source and churned out as a waste product the aggressive oxidizing molecule into the environment. The amount of damage inflicted on oxygen-intolerant species must have been profound, but since we have little to no record of fossilized bacteria, it is hard to estimate.

One clear effect the oxygen release had on the planet was to trigger an ice age. Extending approximately from 2.4 billion to 2.1 billion years ago, this Huronian ice age was allegedly caused by the newly released oxygen into the atmosphere, where it attacked and decomposed methane, a powerful greenhouse gas. The resulting temperature drop added to the stress that oxygen poisoning had already inflicted on the biosphere.

Another ice age stands out before the rise of the animal and plant kingdoms. It is aptly named the Cryogenian period, and it took place 720 million to 630 million years ago, when the near totality of the planet was iced over—not only continents but oceans as well. This "snowball-Earth" episode occurred just before life underwent a dramatic surge in complexity, known as the Cambrian explosion, starting around 540 million years ago, when new body plans led to all groups of animals that would ultimately populate the Earth. Could this ice age be responsible for a "bottleneck" in the course of evolution that pruned the tree of life so radically that it triggered the flourishing of complex new lifeforms in its wake? Scientists are still pondering this intriguing coincidence in timing.

The Late Ordovician Crisis (445 Million Years Ago)

Once silicon-, calcium-, and phosphorus-based outer shells and inner bones came into being, marine animals left a much more durable imprint in the fossil record, making it possible to study

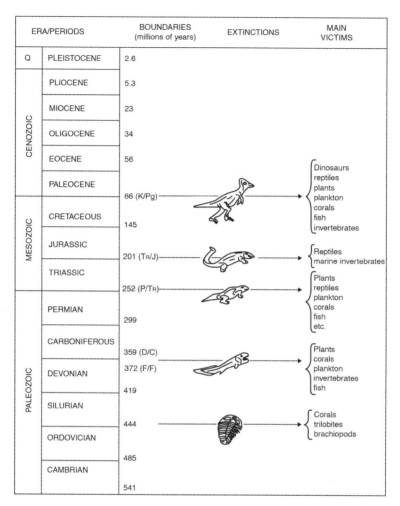

ERA/PERIODS		BOUNDARIES (millions of years)	EXTINCTIONS	MAIN VICTIMS
Q	PLEISTOCENE	2.6		
CENOZOIC	PLIOCENE	5.3		
	MIOCENE	23		
	OLIGOCENE	34		
	EOCENE	56		
	PALEOCENE			Dinosaurs reptiles plants plankton corals fish invertebrates
MESOZOIC	CRETACEOUS	66 (K/Pg) — 145		
	JURASSIC	201 (Tʀ/J) —		Reptiles marine invertebrates
	TRIASSIC	252 (P/Tʀ) —		Plants reptiles plankton corals fish etc.
PALEOZOIC	PERMIAN	299		
	CARBONIFEROUS	359 (D/C)		Plants corals plankton invertebrates fish
	DEVONIAN	372 (F/F) 419		
	SILURIAN	444		Corals trilobites brachiopods
	ORDOVICIAN	485		
	CAMBRIAN	541		

The five great mass extinctions detected in the fossil record and some of their main victims.

in detail the first well-documented mass extinction that rocked the biosphere, 445.6 million years ago, at the end of the Ordovician period.

The crisis is convincingly tied to another ice attack. Traces of glaciers are reported deep inside Africa, scouring the rocks that date back to the period. The damage to the biosphere was particularly severe, because the Earth had previously enjoyed over 100 million years of balmy weather and high sea level, which had spread warm, shallow waters far into the continents and provided a variety of ecological niches for the fast-diversifying lifeforms.

For some mysterious reason, a very rapid cold snap hit the planet, and with the spread of glaciers across the continents, water was pumped out of the oceans, dropping the sea level about 100 meters (330 ft.). Waters retreated from the shallow continental shelves that fringed the deep oceans. With habitats shrinking, increased competition developed between the inhabitants of the continental shelf. An estimated 85 percent of marine species were wiped out: corals, sponges, sea snails and other mollusks, and many trilobites, as well as early armored jawless fish. On land, the wave of extinctions hit early plants and the first ground-dwelling arthropods—the ancestors of centipedes, spiders, and insects.

When the climate warmed again, after approximately 2 million years, the biosphere was shaken up with a new round of losses affecting species accustomed this time to cold water. Finally, after this second hit, the ecosystem recovered, to grow and diversify anew, picking up the course of evolution where it had left off.

The End-Devonian Crisis (372 Million Years Ago)

Over the next 70 million years—a long and stable stretch of time known as the Silurian and Devonian periods—coral reefs made a comeback, hosting new colonies of squid-like mollusks, primitive fish, crawling sea snails, trilobites, and other arthropods, while on land, mosses and treelike plants welcomed spiders and insects, followed by the first vertebrates to crawl out of the waters, which resembled primitive salamanders.

Unfortunately, this rejuvenated ecosystem was hit some 372 million years ago by the second big crisis on record: the end-Devonian mass extinction, which apparently occurred in several pulses. The most severe one, known as the F-F crisis or the Kellwasser event, took place between two well-defined fossil assemblages: the Frasnian and Famennian stages of the Late Devonian. One million years later, a second pulse was recorded at the very boundary of the Devonian and the next period of geological history—the Carboniferous—and is hence referred to as the D-C crisis, also called the Hangenberg event.

There is considerable debate on how protracted and multi-staged this mass extinction really was, and how many species were affected. Bunching together all victims that disappeared over an interval of a couple million years, it does look like 75 percent of species went extinct worldwide, which would indeed qualify as a great mass extinction. Some scholars point out, however, that the only true abrupt crisis within this time frame is the F-F extinction, and that only 40 percent of species were affected at that point.

Be that as it may, the most affected lifeforms were marine animals, with a spectacular crash of coral reefs, accompanied by the extinction of many sponges, trilobites, and primitive fish. Land forests and insects appear to have been much less affected. As is the case for the earlier end-Ordovician mass extinction, the causes of the end-Devonian crisis remain somewhat of a mystery. One clue is that the marine realm appears to have been hit by multiple surges of oxygen-poor deep waters, which spilled onto the continental shelves and choked marine animal life.

During the 1980s, with the realization that asteroid and comet impacts can be devastating, planetologists called attention to the fact that several cosmic collisions did occur at the end of the Devonian period. The Siljan impact crater in Sweden, about 50 kilometers (30 mi.) across, is dated at 377 million years ago, plus or minus 2 million years, which is 4 or 5 million years before the crisis: close but not compelling. Equally intriguing are the Alamo impact beds that stretch over southern Nevada and western Utah, caused by an asteroid strike on a shallow seafloor. The undiscovered crater is thought to have measured anywhere from 50 to 150 kilometers (30 to 90 mi.) in diameter, but its age is again estimated to be 4 million or 5 million years too old to be the cause of the F-F crisis.

The bunching of documented impacts so close to the extinction event, however, could reflect a salvo of projectiles from outer space caused by the breakup of an asteroid or a comet into multiple fragments. In such a scenario, most fragments would have hit the ocean, leaving little physical evidence but flushing deep, oxygen-poor water onto the continental shelves.

An alternate scenario calls for unusually large volcanic eruptions, wreaking havoc in the climate, be it a cold or a warm spell. Supporters of such a volcanic trigger cite extensive volcanism occurring at the time in Siberia (Viluy Trap Basalt province) and on the seafront (Center Hill volcanism). Here again, however, eruption and extinction ages do not coincide precisely, and formal evidence is lacking, despite surges of mercury in the F-F sediment that proponents of the theory attribute to volcanism. Moreover, if great eruptions are to blame, why was life relatively spared on land and more affected in the oceans? A wide-open field of research for students and scholars.

The "Great Dying": The End-Permian Mass Extinction (252 Million Years Ago)

Although one can still harbor doubts about the severity of the end-Devonian crisis, the end-Permian disaster, 252 million years ago, undoubtedly qualifies as the most severe mass extinction of all time. An estimated 95 percent of all species were exterminated, and far from being a predominantly marine crisis, as seems to have been the case at the end of the Devonian, the end-Permian mass extinction whips through oceans and continents alike, cutting down three-quarters of all land vertebrates, including large amphibians and reptiles.

The world that was hit had greatly changed; evolution had introduced many improvements since the previous crisis more than 100 million years prior. On land, the early, small amphibians had diversified and left the safety of riverbanks to probe deeper into the continents, chasing insects for food and developing new reproduction strategies: hard-shelled eggs that retained precious moisture and allowed the animals to survive away from water. This did not prevent all species, including these new high-performance models, from being severely affected by the mass extinction. Researchers agree that the crisis was particularly abrupt. According to the fossil count, its major pulse occurred within a time span of 100,000 years or less.

Something very damaging hit the ecosystem on a global scale. There are no major recognized asteroid impacts that are dated at

252 million years. Major volcanic eruptions, however, did take place in Siberia at the time: one of a number of rare, voluminous outpourings at the Earth's surface, known as continental flood basalts, that make up a large igneous province, or trap.[5] Such events occur every few tens of millions of years on average, but not all are tied to major upheavals of the biosphere. So why was this eruption pulse so devastating, if it is indeed to blame?

Volcanologists point out that the Siberian flood basalts are laced with thick deposits of pulverized ash, which is at odds with the fluid outflows of lava that normally characterize such events. One hypothesis is that the magma rising to the surface struck seams of coal and methane-rich layers of permafrost, triggering explosions and injecting large amounts of greenhouse gases into the atmosphere. If this was indeed the case, a comparison can be made with our current civilization-induced episode of global warming. The end-Permian mass extinction might illustrate an unfortunate convergence of circumstances—volcanism hitting a deposit of fossil fuel—in the same way that, today, civilization-induced global warming could melt the permafrost and release the methane it contains, providing an extra burst of greenhouse gas to the atmosphere.

The End-Triassic Mass Extinction (202 Million Years Ago)

The biosphere took a long time to recover from the "Great Dying" of the end-Permian: 5 million years for animals at the top of the food chain, such as early dinosaurs, and at least 30 million years for the total ecosystem to return to its previous richness and diversity.

Just as it appeared to have recovered, the biosphere took another blow, 202 million years ago, at the end of the Triassic period. Animals and plants seem to have experienced a 75 percent species loss, which would put the end-Triassic event on par with other mass extinctions, although there is much debate as to how long the turnover of species really lasted, how much was due to extinctions as such, and how much was a result of a low creation rate of new species (i.e., a decrease in speciation).

No cosmic catastrophe left its mark 202 million years ago, except for an impact crater, 25 kilometers (15 mi.) wide, in central France, around the village of Rochechouart, with an age estimated at 201 million years in one study and 206 million years in another. The Rochechouart impact certainly had a devastating effect over Europe, but it shows no link to a global upset of the ecosystem.

Volcanism, on the other hand, could have played a significant role in the end-Triassic mass extinction, as it did in the previous end-Permian mass extinction. Continental plates, which had come together millions of years prior in one supercontinent named Gondwana, were on the move again. An overheated zone between North America and Africa began to split, starting the buildup of an ocean floor that would soon become the Central Atlantic. This Central Atlantic magmatic province, or CAMP for short, spewed forth much gas and lava as it reached its peak activity, 201 million years ago, flooding an estimated 2 million square kilometers (800,000 sq. mi.) with hot lava. The large amount of carbon dioxide released into the oceans and belched out into the atmosphere might have been the major cause of the biological shake-up; this was compounded by the fact that the sea level was particularly low at the end of the Triassic period, further fragilizing the marine ecosystem due to the reduced size of the life-bearing continental platforms.

If one were to summarize the first four great mass extinctions in the fossil record, the triggering causes appear to be quite varied: major glaciation at the end of the Ordovician; possible choking of the marine realm by turbid waters and atmospheric upheaval, through volcanism or asteroid impacts, at the end of the Devonian; lethal interaction of a great magma pulse with coal seams or another fossil fuel at the end of the Permian; and volcanism again or disequilibrium between background extinctions and speciation at the end of the Triassic.

Scientists and students of the earth sciences have a lot of exciting research to pursue in order to elucidate these great mysteries of the past. But before we focus on the situation today, and the ongoing wave of extinctions, there is one more major crisis to present—one that merits a full chapter: the extremely abrupt end-Cretaceous mass extinction, which took out the dinosaurs and 75 percent of world species, 66 million years ago. Because it is the most recent of the "Big Five," it is particularly well documented in the fossil record—we have some spectacular "snapshots" of the catastrophe, a variety of extinction mechanisms, and a precise tally of victims and survivors.

2

The End of the Dinosaurs

The great mass extinction that wiped out the dinosaurs 66 million years ago had the profound consequence of promoting the puny mammals that once lived in the shadow of the dinosaurs, a change of cast that ultimately led to the human species. Stated in this way, it would seem that evolution followed a preestablished direction, ultimately leading to the human species as the inevitable outcome. The same line of thought implies that dinosaurs were doomed from the beginning, being much too primitive to get in the way of humankind's manifest destiny. To accommodate this view of evolution, dinosaurs are often portrayed as clumsy animals incapable of adapting to changing environments—to the point where, today, a person or machine without flexibility is called a dinosaur.

During the 1970s, as research progressed in the matter, paleontologists began to challenge this simplistic outlook. Dinosaur fossil bones and footprints span 150 million years of Earth history and represent thousands of different species that flourished on all continents, diversifying and specializing to fill every conceivable niche. Such a global success, over such a long period of time, is proof enough that dinosaurs were able to adapt to changing climates and environments. Therefore, that they ultimately vanished should come as a surprise, and indeed, when paleontologists began to focus on the timeline of the disappearance of dinosaurs, many left their preconceived ideas behind and became convinced that dinosaurs were struck down at the peak of their splendor by some mysterious disaster.

A Global Crisis

The second revelation that caught the attention of paleontologists was that large land-dwelling reptiles were not the only animals that bit the dust during the end-Cretaceous mass extinction. Flying reptiles were also wiped out, the emerging bird group was severely cut back, and even the mammals saw their numbers plummet. The marine realm experienced a collapse of its own, from the plankton world, through countless mollusk and fish species, up to the total wipeout of large marine reptiles, including the 6-meter (20 ft.) plesiosaurs and mosasaurs.

The scope and synchronicity of the end-Cretaceous mass extinction across such different realms was not recognized at first. Dinosaur experts knew little about the fate of marine reptiles and even less about plankton or flowering plants, just as bird specialists today have little knowledge of the plight of elephants in Africa or the collapse of codfish stocks off the coast of Newfoundland.

Focusing on their area of expertise, and unaware of synchronous blows across the entire ecosystem, dinosaur experts came up with a number of explanations to account for the demise of their pet animals: a deadly viral pandemic cutting across the entire family, the popping up of novel flowering plants that poisoned the unsuspecting herbivores, some form of climate stress interfering with their reproductive abilities, or even the robbing of their eggs by malicious mammals challenging the clumsy reptiles.

Seashell and ammonite specialists, for their part, speculated on the debilitating effects of low sea-level stands cutting down the area of life-harboring continental shelves. By the early 1970s, however, specialists in each field began to look sideways at what was happening outside of their realm, and they started to realize that their group of animals was not the only one collapsing at the end of the Cretaceous. They began to seek a global mechanism for what they now understood to be a biosphere-wide trauma. However, there were two different ways to interpret the data.

Dinosaur dig in end-Cretaceous strata, Campagne-sur-Aude (southern France). Photo by author.

One interpretation was rooted in the dogma that ruled the field for the previous two centuries: the belief that changes are slow and progressive across the face of the Earth. A climate upheaval stretching over hundreds of thousands of years, for instance fueled by volcanic eruptions, might also explain the end-Cretaceous turnover as a deadly but progressive phenomenon. Species of dinosaurs vanished one after the other in a prolonged crisis, as did marine reptiles, ammonites, and other victims of the great transition. Many dinosaur experts felt comfortable with this gradualistic perspective since their tally of fossils appeared to indicate a slow decline in the amount and diversity of dinosaur species across several million years.

Another way of looking at the data was to view the progressive decline of species at the end of the Cretaceous as deceptive: a warped perspective due to the scarcity of fossils—especially when it came to large and rare animals—and due to the difficulty of dating them precisely. Paleontologists had access to scattered and often imprecise listings of the last occurrences of species—the sediment level above which their fossils were no longer found—

and plotting them often produced a graph showing "stepwise," offset extinctions of various species at different times. There was a belief that if geologists kept poring over the sedimentary record, they would ultimately discover new fossils, the last occurrences of which would progressively converge toward the same exact level and prove that the end-Cretaceous crisis had been brief and brutal. Proponents of such a catastrophic scenario also based their hunch on the fact that the end-Cretaceous extinctions affected such a variety of species, marine and land-based alike, that the trigger had to be unusual. An exceptional crisis called for an exceptional cause, but what on Earth might that be?

The Answer Is in the Clay

One way of solving the issue was to find, somewhere on Earth, layers of sediment that were sufficiently fine and complete across the interval of the crisis to display the precise sequence of events. Shallow marine sediments are particularly well suited for the job: their plankton fossils, plentiful and covering many individual species, offer a high-resolution reading of events.

There are several marine outcrops in Europe, of Late Cretaceous age, that are uplifted above sea level and easily accessible, in particular one stack of plankton-rich limestone in the Italian Apennine mountain range, near the medieval city of Gubbio, and another fine outcrop of chalky limestone exposed in a cliff face at Stevns Klint in Denmark, south of Copenhagen.

In the 1970s, as they were combing through these marine sediments and tallying plankton species, paleontologists identified a thin layer of clay, less than a couple of centimeters (1 in.) thick, that appeared to show an abrupt turning point in plankton history. Below the rust-colored clay, a diverse set of plankton species populated the end-Cretaceous seas. Above the layer only a couple of dwarf species remained. In Denmark, the clay itself was conspicuously rich in fossil fish remains—to the point where it became known as the "fish clay"—as if it were some sort of marine cemetery.

In land ecosystems, wherever lacustrine or river sediments

At Raton Basin (southern Colorado), the K-Pg boundary takes the form of a 2.5-centimeter (1 in.) layer of light-colored clay (below the black coal seam). Photo by Kirk Johnson, Denver Museum of Nature & Science, courtesy of National Science Foundation.

preserved a high-resolution record of the end-Cretaceous biota in the form of fossil plant seeds and spores, a similar pattern emerged. Here, as well, the diversity of plant life typical of the Late Cretaceous period was sharply interrupted by a similar layer of clay and replaced above it by an impoverished collection of only a few fern species.

Ubiquitous across all sites around the world, wherever the end-Cretaceous record was preserved, the thin layer of clay became known as the K-T boundary layer, marking the sharp frontier between the Cretaceous period (symbolized by the letter K in geological nomenclature) and the Tertiary era of Earth history (symbolized by the letter T). The name "K-T" had a good ring to it, but the nomenclature was changed in 2013 to "K-Pg" to use the better-accepted Paleogene period (Pg) in place of the obsolete Tertiary denomination (T). It is therefore the K-Pg symbol I will now use in this book.

Not only did the end-Cretaceous K-Pg crisis begin to look razor-sharp in the late 1970s—at least as far as plankton and plants were concerned—but the clay itself looked as if it might shed light on the nature, and perhaps even the duration, of the worldwide trauma. Clay is a deposit of fine mineral particles leached off the land and discharged by rivers and wind into lakes and seas. In peaceful, relatively deep marine settings, as were the Gubbio, Italy, and Stevns Klint, Denmark, sites at the time, a layer of clay a couple of centimeters (1 in.) thick was expected to take anywhere from 1,000 to 10,000 years to collect on the sea bottom. But since this particular clay betrayed some exceptional process, perhaps it formed much faster. Perhaps it even contained chemical or mineral clues as to the life-killing process itself.

One American geologist in particular, Walter Alvarez, decided to pick up the challenge after coming across the mysterious clay layer during his fieldwork in Gubbio. Upon returning to Berkeley, California, he convinced his father, Nobel Prize–winning physicist Luis Alvarez, to come up with a system to estimate what time-span the clay might cover. The idea Luis Alvarez came up with was to measure the amount of micrometeorites in the clay.

Micrometeorites are cosmic dust particles from asteroids and comets that rain down from the sky at a fairly regular rate. The higher their concentration in the clay, the longer that clay took to form on the sea bottom as it amassed the infalling particles. Rather than count micrometeorites under the microscope—a time-consuming and boring ordeal—the team used a proxy instead: the concentration of iridium, a platinum-class metal that is excessively rare on Earth but found in small amounts in micrometeorites. The technique was all the more convenient because they could make the iridium radioactive by bombarding the clay sample with neutrons; gamma rays emitted by the activated iridium could then be counted automatically.

Geochemist Frank Asaro joined the Alvarez team to make these very precise measurements. Indeed, the average concentration of iridium is little more than 0.0001 percent in micrometeorites—1,000 parts per billion, or 1,000 ppb—so that after being diluted in the sea-bottom clay, the iridium concentration was expected to range somewhere between 0.01 and 0.1 ppb, depending on how long the clay took to form, which was the answer sought.

The investigators were in for a big surprise. The iridium concentration in the K-Pg clay turned out to be astonishingly high: over 10 ppb—in other words, one hundred to one thousand times higher than what they expected. It was as if a mere inch of clay had taken a million years to build up on the sea floor, during which time the oceans had been devoid of life, since the clay contained no plankton. Clearly, something was wrong with the cosmic hour-glass method.

One explanation was that 66 million years ago a giant star had exploded close to the solar system, showering the Earth with rare chemical elements, like iridium, as well as life-threatening gamma rays. Indeed, scientists who had called attention to mass extinctions in the first place had suggested star bursts to explain the scope and severity of such global crises. In 1963, Otto Schinde-wolf had precisely suggested a supernova to account for the end-Permian mass extinction, and a few years later, Canadian dinosaur expert Dale Russell had suggested an exploding star to explain the end-Cretaceous crisis.

If this was the case, the K-Pg clay should be laced with other heavy metals that could only be produced in supernova explo-sions, such as plutonium-244. Plutonium expert Helen Michel stepped in to perform the measurements but came up with a neg-ative answer: there was no plutonium-244 in the boundary clay. Exit the supernova theory.

Running out of ideas, Luis and Walter Alvarez spent months pondering their mysterious results. With hindsight, it is easy to-day to jump to the conclusion that the impact of an asteroid—a cosmic body containing the requisite amount of iridium—was the logical answer to the problem. Back in the 1970s, however, large impacts on Earth were seldom discussed among geoscientists, let alone taught, and if the thought did cross their minds, Luis and Walter Alvarez failed to recognize how an impact somewhere on Earth could affect the entire biosphere.

The eureka moment came when Luis Alvarez made the con-nection with a seemingly unrelated field: the great volcanic erup-tions of the past. He recalled how the Krakatau eruption of 1883 and especially the Tambora eruption of 1815, both in Indonesia, triggered a global cooling of the Earth's atmosphere: nearly 1°C (1.8°F) over the course of an entire year, because of dust particles and sulfate crystals that spread around the globe, reflecting sunlight back to space. Could it be that an asteroid impact had set off a similar but much larger perturbation, lofting such an opaque veil of dust that it plunged the Earth's surface into an endless and chilling night for weeks or months on end? Was photosynthesis completely shut down on land and in the oceans, leading to the complete breakdown of the food chain and the choking and starvation of most living species? Armed with a credible mechanism to explain the devastating aftermath of an asteroid impact, in June 1980, Walter and Luis Alvarez, with Frank Asaro and Helen Michel as coauthors, were ready to publish their analysis of the K-Pg clay in the American journal *Science*.[1]

The article made quite an impact, in every sense of the word. It challenged the ruling opinion among geologists that Earth processes were necessarily slow and progressive. It was disturbing, but it had to be taken seriously because it followed the golden principles of a sound scientific theory: it was based on detailed fieldwork and data analysis in the lab, and it led to a set of predictions—a number of research leads that could help validate or refute the theory. The article suggested that if its premises were right, one should find more clues pertaining to an asteroid collision in the K-Pg clay, as well as a 66-million-year-old impact crater, somewhere on Earth. The authors even went as far as to calculate the size of the putative crater. Given the amount of iridium dispersed around the globe and knowing the average iridium content of an asteroid based on meteorite samples, a simple equation pinned down the mass and the volume of the impactor: roughly 10 kilometers (6 mi.) in diameter—a body the size of Mount Everest. Another equation related the mass of the impactor to the size of the crater it would produce on Earth: a basin roughly 200 kilometers (125 mi.) in diameter. At first glance, however, there was no crater of such size and of the right age at the surface of the Earth.

A Volcanic Alternative

In the absence of a culprit crater, it was not long before geologists came up with another theory to explain the high iridium content of the K-Pg clay. In 1983, paleontologist Dewey McLean pointed out that large and prolonged volcanic eruptions had taken place in India at the end of the Cretaceous period—and that they might have released enough carbon dioxide into the atmosphere to trigger an episode of global warming that overwhelmed the biosphere. It is worth noting that McLean's intuition, regardless of how well it held up in the debate pertaining to the end-Cretaceous mass extinction, was a prescient warning of the effects that global warming might have on our biosphere today.

Other scientists joined McLean in promoting the volcanic scenario as a preferable, more conservative alternative to the cosmic impact theory, since it relied on familiar Earth processes, rather than on an exotic new agent, and it lent support to the idea of a protracted decline of the biosphere, rather than a brutal one, which was more in keeping with traditional views of evolution. Such support for volcanoes causing the K-Pg crisis came from geologists Charles Officer and Charles Drake in the United States and Vincent Courtillot in France, among others.[2]

The incriminated lava fields in India, known as the Deccan Traps, were extensive indeed, and their timing did overlap the end-Cretaceous biotic crisis. Such massive outpourings of lava occur every few tens of millions of years on average. The latest one, known as the Columbia River Basalts, erupted in the American Northwest 16 million years ago and covered 210,000 square kilometers (81,000 square mi.), the area of Idaho. Further back in time, a lava field three times larger was emplaced in Ethiopia around 30 million years ago. These voluminous, hot, fluid flood basalts are by no means created instantaneously: their eruptions are spread out over hundreds of thousands, if not millions, of years. They consist of intermittent eruptions—each spewing forth several hundred cubic kilometers of lava—separated by rest periods lasting centuries to millennia.

With respect to the end-Cretaceous crisis, supporters of the volcano hypothesis suggested that the K-Pg clay represented the fallout of such an eruption in India that blew particles around the globe, including the anomalous iridium. In their scenario, the rare metal did not come from some rogue asteroid but from deep inside the Earth's mantle, close to the core boundary, pumped up to the surface by a hot-spot plume. The concept was worth testing, but the results were unconvincing. Chemical analyses of Deccan basalts showed that even if the entire lava field had released the near totality of its iridium into the atmosphere, it would amount to less than 1 percent of the amount present in the K-Pg clay. The meteoritic origin of the iridium was further confirmed by measuring other rare metals present in the clay: they showed proportions typical of material from outer space, distinctly different from those found in terrestrial lavas.

Throughout the 1980s, new clues kept pouring in, favoring the impact theory over the volcanic one. Peculiar droplets of glass, recovered from the clay, were analyzed by Dutch paleontologist Jan Smit. Their composition and radial texture showed that they were not droplets of lava but vaporized rock that condensed out of a fireball in the vacuum of space. In 1983, Bruce Bohor of the US Geological Survey discovered quartz grains in the K-Pg clay that were shocked and deformed by exceptionally high pressures that were only compatible with asteroid impacts and nuclear blasts. The volcanic camp counterargued that spherules and deformed quartz grains could be caused by volcanic explosions, but those claims did not hold up to further scrutiny.[3]

When a theory relies on a number of assumptions—in this case that the iridium, glass spherules, and shocked quartz grains were volcanic—and when all these assumptions are proven false, one might expect the theory to collapse. Proponents of the volcanic scenario did not give up their research, however, and shifted their argumentation away from trying to explain the characteristics of the K-Pg clay. They fell back onto the claim that the mass extinction was progressive, rather than sudden, and that protracted volcanic eruptions fit the picture better than a sudden impact.

The Crater Uncovered

Throughout the 1980s, proof of a giant impact was piling up in the K-Pg clay, but the purported crater was still missing. There were ways to explain the discrepancy: during the 66 million years

The Shifting Role of Volcanism

The role of volcanism in the end-Cretaceous mass extinction shifted many times since it was advocated in the early 1980s. It was first suggested that eruptions in India—the Deccan Traps—had reached a climax that deposited the worldwide K-Pg boundary clay and was responsible for its iridium and shocked minerals, negating the very occurrence of an asteroid impact: volcanism alone had caused the crash of the biosphere. This early hypothesis was proven false when the clay was conclusively tied to an asteroid impact and was found to hold no volcanic component.

The volcanic hypothesis was then amended: while it was conceded that an asteroid impact had occurred and was the source of the K-Pg clay, it was stated that eruptions had weakened the biosphere *before* the impact, playing just as big a role, if not greater, in the severity of the ensuing crisis. The argument relied in part on the assumption that the eruptions were particularly severe over tens of thousands of years leading up to the impact. Although this amended theory waned over the years because evidence was lacking, it was reintroduced in 2022, based on new field studies in northern China.[4]

Another twist to the volcanic theory took place in 2015, when a careful analysis of India's lava fields showed that the bulk of the eruptions might have taken place *after* the impact, rather than before it. A new amendment to the theory claimed that while the eruptions no longer played a role in the demise of the biosphere, they were crucial in slowing down its *recovery*.[5] There is no firm evidence yet to support this claim, and climate modeling of Deccan volcanism hints that it is incapable of significantly harming the biosphere or delaying its recovery but might on the contrary have had a positive effect, mitigating through mild greenhouse warming any cooling effect the impact might have had—and hastening the recovery of surviving species.

since the collision, the crater might have been erased by erosion or else buried under layers of sediment. Considering the areal extent of oceans on Earth, chances were also high that the asteroid hit the ocean floor, where the crater would be much harder to detect, or was even swept away by seafloor spreading into a trench, crumpled and melted back into the Earth's mantle. If this had been the case, however, an oceanic impact would have left somewhere along the coastline the tell-tale traces of a giant tsunami.

Geologists in search of the crater held on to another hope. If the K-Pg clay represented the fireball expelled from the impact site, then the thickness of the fallout layer had to increase substantially toward ground zero, pointing in the direction of the crater. Judging from such ejecta layers around other impact craters on Earth, and extrapolating to the 200-kilometer (125 mi.) diameter of the missing K-Pg candidate, its layer of debris was calculated to reach a thickness of 50 centimeters (20 in.) at a distance of 1,000 kilometers (620 mi.) from ground zero, and over 5 meters (16 ft.) within 500 kilometers (310 mi.) of the crater.

Such hopes were reinforced by the apparent doubling of the K-Pg clay thickness in North America: up to 2 centimeters (close to an inch), compared to the thickness of 1 centimeter (less than half an inch) on European sites. The breakthrough came when Canadian PhD student Alan Hildebrand discovered, on the banks of the Brazos River, near Houston, a layer tens of centimeters (over a foot) in thickness, packed with glass spherules and spiked with iridium. Equally significant was a sand bed, 1 meter (3.3 ft.) thick, lying just above the boundary layer: it pointed to violent shaking of the seafloor at the time, leading to underwater slumping of sandy sediment, or some sort of tsunami. Reviewing the scientific literature pertaining to other end-Cretaceous sites around the Gulf, Alan Hildebrand noted a report signaling another instance of a thick layer of spherules, this time on the island of Haiti, described a few years earlier by Haitian geologist Florentin Maurasse. A quick fieldtrip to the island convinced Hildebrand and his coworker David Kring that this 50-centimeter (20 in.) spherule bed was another occurrence of the impact's ejecta layer. Then came in more reports of spherule beds and iridium, tens of centimeters thick, inside Mexico, close to the coast, and toward the Atlantic, in a marine core. With several data points, including two that were theoretically some 500–1,000 kilometers (300–600 mi.) from ground zero—the Haitian and Texan ones—and one more data point at sea, it was now possible to locate the crater by triangulation; that is, by tracing circles on a map, radiating from the data points. Their intersections pointed to a location in the Gulf of Mexico, just north of the Yucatán peninsula. No impact crater stood out in the area, however, on geographic maps or satellite imagery. But on a map of the Earth's gravity field, recording subtle density variations below the surface, a circular anomaly roughly 180 kilometers (110 mi.) in diameter did appear, underlying the northern shore of Mexico's Yucatán peninsula.[6]

This is where the story of the missing crater takes on a surprising spin, worthy of a mystery novel. Yucatán's circular structure had been discovered shortly after World War II during a gravity survey, and it was reached by drilling in the 1950s, with the hope that it was a buried sedimentary basin containing petroleum. Pemex, the Mexican oil company, drilled down over 1 kilometer (more than half a mile) until it penetrated what it believed to be lava layers, based on the cores brought back to the surface, which dashed any hopes that the structure might contain oil or gas.

The state-owned company did commission a new aerial survey in the 1970s, this time to collect magnetic data across the Yucatán. An American consultant on the team, Glen Penfield, happened to be familiar with impact craters. When Penfield plotted the data, he saw a huge circular structure emerge, 180 kilometers (110 mi.) across. Learning from his Mexican colleagues that it matched a gravity structure previously identified at the same location, he became convinced that he had tripped on a buried impact crater. By then, it was common knowledge that scientists were searching for an impact crater to explain the extinction of the dinosaurs, and Penfield knew he had a winner. He convinced his boss, Pemex geologist Antonio Camargo, to share their discovery with their colleagues at a symposium of petroleum geology in Los Angeles in October 1981 and announce that the feature might well be the culprit of the end-Cretaceous crisis.

The cat was out of the bag. Journalist Carlos Byars of the *Houston Chronicle* picked up the story. His article made the front page of the Texan newspaper on December 13, 1981, titled "Mexican Site May Be Link to Dinosaurs' Disappearance," but, incredibly, it went unnoticed, at least by professional geologists. For years, Carlos Byars kept mentioning the hypothetical crater at science conferences, trying to attract attention to Penfield and Camargo's findings. No one seemed to pay attention, perhaps because the information came from an outsider who did not belong to the scientific establishment.[7]

It wasn't until 1990 that the news about the candidate crater finally reached Alan Hildebrand, who then tracked down Glen

Penfield. Realizing that each held different pieces of the puzzle, the pair joined forces to pursue the quest together and managed to recover old drill samples of the Yucatán underground structure. They immediately recognized the cored rock fragments to be impact melt, not volcanic lava, and with this final proof in hand, set out to publish their findings.

There was still one obstacle to overcome: the judgment of peers. In a science journal, it is customary for an article to be reviewed by a panel of experts who might suggest improvements or even reject the article for being insufficiently researched or unconvincing. When Hildebrand, Penfield, and their coauthors submitted their paper to *Nature*—a prestigious and widely read science journal—two referees out of three rejected it, perhaps because the referees supported the work of other scientists and were convinced that the crater lay elsewhere. In science, fortunately, the truth always manages to come out. A few months after its rejection by *Nature*, the article was published instead by *Geology*, an American science journal, on September 1, 1991.[8] It made as

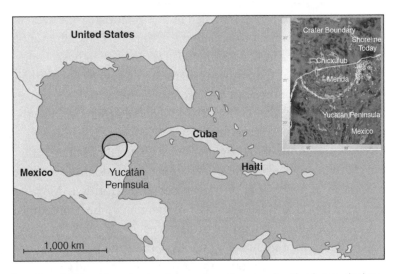

Location of Chicxulub crater. *Inset,* map of gravity anomalies that betray the location of the crater, over 1 kilometer (0.6 mi.) underground. White dots represent sinkholes (called cenotes), created by the dissolution of limestone by groundwater: a circular concentration of cenotes underscores the fault line that marks the periphery of the crater. Gravity map by Geological Survey of Canada.

much of an impact as the Alvarez article, 11 years prior: not only had the doomsday crater been discovered at last, but it perfectly fit the expected profile.

Portrait of a Killer

The crater was named Chicxulub (pronounced *Chik-shoe-loobe*), the Mayan name of the village closest to ground zero, near the Gulf of Mexico coastline. At the time of the impact, 66 million years ago, the global sea level was higher than it is today, and the Yucatán was a sea-flooded basin, deepening from roughly 100 meters (330 ft.) in the southwest to over 1,500 meters (1 mi.) in the northeast. The violent impact changed the layout of the region in a matter of minutes, digging a deep hole almost down to the Earth's mantle. The slopes of the steaming hole collapsed instantly; huge terraces of rock, streaming with molten lava, slid back into the cavity. On the periphery, blocks of upturned rock, showered by the curtain of ejecta falling back from the sky, built up a rim that briefly peaked above water before landslides and strong currents tore it apart, letting the ocean waters flood the crater. Meanwhile, a salvo of huge tsunami waves rippled outward.

The shaking lasted hours to days, as landslides and tidal waves continued to rock the crater while it progressively settled to reach its final diameter of 180 kilometers (110 mi.). Then, with the passing of time, as peace returned to the area and to the planet as a whole, plankton and sea creatures flourished anew, filling the crater with limestone over millions of years. Bending under its load of sediment, the crater slowly vanished underground, until it reached its present depth of 1,500 meters (1 mi.) below the Earth's surface. Although Chicxulub crater is now out of sight, it can be studied by deep coring, gravity and magnetic surveys, and computer modeling, and by comparison with visible craters of similar size on the Moon and Mars.

At the time of impact, according to calculations, the initial cavity measured 90 kilometers (56 mi.) in diameter and probably up to 20 kilometers (12 mi.) in depth—ten times the depth of Grand Canyon—before rapidly collapsing and widening into a shallow bowl 180 kilometers (110 mi.) wide: an area the size of Vermont or New Jersey. One way to visualize its morphology is to look at the most recent impact craters on the Moon, such as Copernicus or Tycho, with a pair of binoculars or a small telescope. Such craters are surrounded on the outside by a rim of ejecta, thinning outward, and display slumped terraces on the inside that lead down to a relatively flat floor. In the center rises a positive relief: a central peak for craters of the size of Copernicus or Tycho, and for the larger Chicxulub structure, a central plateau 40 kilometers (25 mi.) wide, resulting from the elastic rebound of the compressed crust.[9]

By the time the collapsing, uplifting, and shaking ended at Chicxulub, the downdrop to the crater floor had adjusted to a mere 2 kilometers or so (about one mile), its shallow bowl filled with slumped material and fallback ejecta that rained down from the sky. Some of the crater fill was partially or even completely molten, resembling volcanic lava—geologists call this suevite, or impact melt—while the bulk consisted of shattered and mixed chunks of target rock glued together by the intense heat into a sort of "rock pudding" called impact breccia.

Those were the rocks that were brought back to the surface by the Pemex exploratory drilling in the 1950s. Chemical analyses showed them to be calcium-rich, matching the composition of the glassy spherules found worldwide in the K-Pg clay. As for the dating of the impact melt, it also yielded the same, precise age as the K-Pg clay: 66.04 million years, plus or minus 10,000 years.[10] Work published in 1992 demonstrating the matching composition and age of Chicxulub's melt rock with the worldwide spherules proved beyond doubt that the Mexican crater was the source of the K-Pg clay around the globe.[11] It validated the Alvarez theory that a cosmic impact was the sole cause—or at least the main cause—of the end-Cretaceous mass extinction.

Not all scientists were convinced, however, especially those who still argued that the mass extinction had been progressive, rather than brutal, and held on to the belief that volcanic eruptions in India played a major role. The controversy kept raging throughout the 1990s, and there are still critics of the impact theory today. This is worth a brief digression, because such controversies illustrate the positive role of critical thinking in science, driving protagonists to sharpen their arguments and collect new data to make their point. On the downside, they also show how scientific research and findings can sometimes be slowed down or blurred by political or personal agendas. Such interference is merely academic when it comes to dinosaurs, but the stakes and potential damage are much higher when it comes to climate change or the reality of present-day extinctions.

Adjusting to a New Theory

Letting go of ingrained beliefs and accepting a new theory can be disturbing, especially for scholars who invested a great deal of their time and put their reputation on the line, betting on the wrong horse. The discovery of Chicxulub crater was not good news for proponents of the volcano scenario. At first, some continued to believe that Chicxulub was a gigantic volcanic caldera rather than an impact crater. Once there was no doubt as to its nature, opponents challenged the age of the feature, arguing that it was significantly older than the K-Pg crisis and therefore could not be its cause. They based their claim on older sediment that had slumped into the crater, but they were simply ignoring the fact that impacts shuffle and shift around preexisting rock layers.

The purported role of the eruptions in India also kept shifting. Although their age did overlap the K-Pg crisis, it was now clear that the lavas bore no physical relationship with the boundary clay. Proponents kept arguing, nonetheless, that the eruptions had provided the major mechanism of the mass extinction, and that the impact was only the finishing blow (see "The Shifting Role of Volcanism"). Blending both impact and volcanism to explain the K-Pg crisis appeared to offer a reasonable compromise, so that no one would lose face.

Science, however, leaves little room for compromise. The volcanic scenario would make sense only if it could be proven that the biosphere declined progressively throughout the duration of the drawn-out eruptions in India, rather than abruptly collapsed at the precise level of the impact layer. As long as paleontologists relied on dinosaur bones and other fossils of large animals—which were rare and often petered out short of the K-Pg layer[12]—the extinctions of various species could be mistaken to be spread out in time (as we saw earlier), and to have started long before the impact. However, by focusing on microscopic species, such as plankton, that leave abundant fossils, paleontologists could build a detailed timescale that much better measured the onset and duration of the biological crisis.

But reading even the plankton record proved tricky: experts still ended up split into two camps. Paleontologist Gerta Keller and others saw different plankton species disappear at different levels in the sediment up to and across the K-Pg boundary: staggered extinctions, spanning 100,000 years or more, as would befit a drawn-out volcanic scenario. Others, including paleontologist Jan Smit, tallied the very same plankton species right up to the clay boundary and vouched that they all disappeared together at that precise level: the moment of impact. Obviously, someone had to be wrong.

In order to resolve the issue, paleontologists decided to stage a blind test. Half a dozen specialists, representing both camps, were given the same set of samples, collected at different levels in a particularly good outcrop of plankton-rich marine sediment—the El Kef site of Tunisia—that spanned the K-Pg boundary. They were not told which sample came from which level, so as not to be influenced in their search and identification of the various plankton species.

The results were enlightening. For any given level, experts came up with different tallies, some failing to spot a species in the upper levels of the sequence and believing that it had gone extinct earlier than it really had. However, when all "score cards" were pooled together, it appeared that *all* plankton species were identified by at least one expert up to the K-Pg boundary: proof that none had vanished prematurely, before the moment of impact. In other words, there was no proof of volcanoes tampering with the biosphere in the years leading up to the asteroid collision.

One conclusion that can be drawn from this blind test, and

from the K-Pg controversy as a whole, is that scientific research is not infallible. In this example, it was only human to miss or confuse a few plankton species. Another point that can be made is that espousing a theory often blurs one's objectivity: it is easy to miss something if one strongly believes that it isn't there in the first place.

One might also wonder why false claims and shaky theories take so long to debunk. Why is it so, more than 30 years after the discovery of Chicxulub crater and demonstrations like the El Kef blind test, that a dwindling group of experts still believes that volcanic eruptions in India played a significant role in the end-Cretaceous mass extinction? One aspect, quite understandably, is personal pride: no one likes to be proven wrong, especially since the job of a researcher is to uncover the truth. There also can be private interests at stake: challenging a popular theory might put authors in the limelight, attract research grants, sell more books, or open doors to radio and TV talk shows. The media often plays along because controversial debates and mysteries are much more exciting to the public than a closed case.

The demise of the dinosaurs and the controversy surrounding its cause might seem a trivial concern in the face of the major crises and challenges faced by our civilization today. But there are lessons to be learned and similar pitfalls to be avoided. Figuring out the correct sequence of events leading up to a crisis and projecting its consequences are skills that need to be honed: enigmas of the past are instructive and precious in this respect. A good example, developed in the next section, consists of figuring out what lethal mechanisms toppled the biosphere in the aftermath of the end-Cretaceous impact and how much interaction and which feedback loops occurred between them, possibly amplifying the crisis.

Scenario of a Disaster: Regional Effects

"Scenario of a disaster" encapsulates rather well the discussion to follow. What took place was indeed a disaster in the true meaning of the word: "the fall of a celestial body," if we go back to its Latin roots. And a scenario is indeed all we can come up with: while the occurrence of the impact is a true scientific theory backed up by evidence, the mechanisms that led to the collapse of the biosphere are much more hypothetical and rely for the most part on educated guesses and computations.

Such efforts to reconstruct the course of events are nonetheless essential, in that some of the mechanisms modeled and described in the end-Cretaceous scenario might also play a role in the wave of extinctions striking our biosphere today. One important area of research is "snowball effects": the synergy that can consolidate and amplify lethal mechanisms beyond their individual contributions.

Before we address the complexity of such mechanisms, it is important to stress the sheer brutality of the impact that triggered the end-Cretaceous mass extinction: a tremendous amount of energy delivered in an extremely brief interval of time. The size of the impacting asteroid or comet is constrained by the amount of iridium dispersed around the globe and by the size of the crater. One set of solutions is that the asteroid's diameter was around 10 kilometers (6 mi.), the size of a mountain, and its velocity around 20 kilometers per second (12 mi. per second), the average speed of an incoming asteroid, relative to Earth.

The kinetic energy of the collision, as physics students will tell you, is proportional to the mass of the projectile multiplied by its velocity squared. If we plug in the numbers and translate the result into familiar units, the end-Cretaceous impact delivered an amount of energy equivalent to 100 million megatons of TNT, which is approximately *ten thousand times* the nuclear arsenal of the entire world, or to call up another gruesome comparison, *six billion times* the Hiroshima bomb.

This tremendous amount of energy was initially delivered to one point on Earth—ground zero, at the northern end of the Yucatán platform—but it then spread to the entire globe. In the first few minutes, the region most devastated was, of course, ground zero. A jet of volatilized rock at temperatures over 10,000°C (18,000°F)[13] spewed skyward and spread out in all directions. Within a radius of 1,500 kilometers (1,000 mi.), about three times the area of Mexico, the fireball probably ignited forests as if they were match sticks. A shock wave rippled through the atmosphere, generating winds that were close to supersonic in speed over a comparable area. As for the ground shaking, it would have ranked a whopping 12 on the Richter scale — one hundred to one thousand times more brutal than the largest earthquakes on record.

Since the impact took place in open water, it triggered a monstrous tsunami. According to computer simulations, the blast drove a wall of rocks outward in all directions, pushing up a wave of water close to 5 kilometers (3 mi.) high. Ten minutes after impact, the amplitude of the rippling wave, traveling across the ocean, was still close to 1.5 kilometers (1 mi.). After four hours, when it reached the American Gulf coast to the north and South America's coastline to the south, it had abated to a swell about 100 meters (330 ft.) high, but this was then amplified manyfold by the shallowing depth as it neared the coast — a common feature of breakers and tsunamis — so that a wall of water over 300 meters (1,000 ft.) tall crashed onto shore. Even as far north as the Canadian Atlantic coast, the swell still had an amplitude of 10 meters (30 ft.), and halfway around the globe, on Australia's and China's eastern coasts, the water run-up was still equivalent to the blow inflicted by the deadly 2004 Indian Ocean tsunami on its closest shores.[14]

What proof do we have to confirm such models? The effects of such a tsunami should be visible on the scoured sea bottom, as well as on land. As soon as they realized that the impact and its tsunami hit the Gulf of Mexico, geologists combed the end-Cretaceous strata in the area, through Texas and into the Mexican state of Nuevo León, where layers of sediment that were under water at the time have since been uplifted by tectonic movements and are open to scrutiny on land. It was not long before geologist Walter Alvarez and paleontologist Jan Smit stumbled upon a layer of sandstone, 3 meters (10 ft.) thick, that sharply interrupted the normal sequence of seafloor sediments. Unmistakably, this was the mark of a giant tsunami that had washed up on shore and drained back a huge volume of sand and debris down the continental shelf.[15]

I was fortunate to visit the Mexican sites with Jan Smit and other geologists in February of 1994, three years after their discovery. Trekking up Arroyo el Mimbral, a dry riverbed, we reached one of the most spectacular outcrops. It was carved by the river into a steep bank and showed the whole succession of layers in cross section: at the base lay a bench of impact spherules, close to a meter (3.3 ft.) thick, that were dropped by the fireball and had settled on the sea bottom; then came 2 meters (6.5 ft.) of sand that flowed over the spherules minutes later, when the tsunami shook up the continental shelf. Fragments of fossilized vegetation, ripped from the coastline and flushed down slope by the tidal wave, were also visible to the naked eye, incorporated into the sand stack. At the top of the pile, the final layers showed a wavy "cross-bed" pattern, freezing in place for posterity the switching directions of the current, as the deep waters oscillated back and forth in the aftermath of the impact.

Driving 200 kilometers (125 mi.) to the north, we found the same tsunami sequence on the site of El Peñon in Nuevo León, thicker yet—a whopping 7 meters (23 ft.) of sand—and likewise topped by undulating layers, with the extra thrill that in some places they were stripped bare by erosion and formed the very ground on which we tread: a strange and moving experience to walk the bumpy bottom of the seafloor as it was shaped on doomsday, some 800 kilometers (500 mi.) from ground zero.

Disaster Scenario: The Big Picture

One can easily imagine how large populations of animals and plants were massacred within a short range of ground zero, drowned by the tsunami or incinerated by the heat wave. It is less obvious why the entire biosphere was struck worldwide, to the point where countless species went extinct.

When the impact theory was published in 1980 to explain the mass extinction, Luis Alvarez suggested that the dust load injected into the atmosphere was enough to darken the entire globe, shutting down photosynthesis and causing the whole food chain to collapse. The concept relies on the enormous amount of vapor-

At the K-Pg boundary on the Tanis site of North Dakota, Robert DePalma and his team struck a tsunami deposit containing fossilized sturgeon and paddlefish choked by spherules from the asteroid impact. Photo by NASA / Taylor Mickal.

Tsunami sand beds at the K-Pg boundary in northern Mexico. Physicist Éric Robin points to the top of the sequence, rich in iridium. Photo by author.

Where Are the Bone Beds?

One would imagine that the massacre of the biosphere in the wake of the asteroid impact would have left obvious dinosaur bone beds and other mass graves, but in fact, the probability of that happening is low. Good preservation conditions must occur for fossils to exist in the first place, and after millions of years of burial, the layers containing them must be unearthed precisely today for us to find.

Luckily, there is one site in North Dakota that offers an extraordinary snapshot of the massacre. The Tanis site—part of the well-studied end-Cretaceous Hell Creek Formation—lies about 3,000 kilometers (2,000 miles) from ground zero in the Gulf of Mexico, but it was reached within tens of minutes by the speedy tidal wave that moved up the Western Interior Seaway of North America and up the rivers that bordered it. Remains of both saltwater and river fish (sturgeons, paddlefish, and bowfins) are found in the sediment layers at Tanis, often in positions typical of suffocation, their gills stuffed with microtektites, the tiny spherules of molten rock that were ejected by the Chicxulub blast. These burial layers, unmistakably contemporaneous with the impact, also bear dinosaur feathers, dinosaur bones, and even a pterosaur (flying reptile) embryo fossil—and one very complete specimen of a Thescelosaurus dinosaur, a bipedal herbivore. A layer of clay immediately below the "killing" level also records footprints of a variety of dinosaurs—triceratops, hadrosaurs, and carnivorous therapods—that confirm that rich dinosaur communities flourished right up to the moment of the lethal asteroid impact. The height of the tidal wave, as it crashed through the Tanis site, is estimated to have been on the order of 10 meters (30 feet).

Another precious detail revealed by the site is that the fossilized paddlefish offer us a hint of the season of year when the deadly asteroid hit the planet. According to experts who studied the fossilized teeth and fins of the paddlefish, which show under the microscope a fine record of seasonal growing cycles, the time of death appears to be at the beginning of the growth season, so during springtime in the Northern Hemisphere.[16]

ized, molten, and pulverized rock launched skyward from ground zero. Calculations linked to other craters on the Earth, the Moon, and Mars show that the rock mass excavated by an impact reaches up to three hundred times the mass of the incoming body. Hence, the material ejected from the Yucatán platform was on the order of 150,000 cubic kilometers (36,000 cu. mi.), comparable to the Big Island of Hawaii down to the seafloor being thrown up into the atmosphere.

This ejecta flew off in all directions from ground zero at speeds comparable to that of the incoming asteroid itself, in the range of several kilometers per second. Such a massive launch was all the more efficient because the atmosphere had been blown away for the occasion, and expanding gases from the volatilized target rock even served as a supplementary booster to accelerate the debris skyward. The shooting rocks followed ballistic trajectories, rising high above the Earth. Some pulled away into deep space; most wrapped around the planet to fall back down like missiles, hitting every area of the globe and spreading dust across the atmosphere worldwide.

Luis Alvarez reached his theory about the impact's lethal effects—the darkening of the skies and the shutdown of photosynthesis—by ramping up the observed consequences of great volcanic eruptions of the past, notably Tambora and Krakatau in the nineteenth century. The physicist was also influenced by the novel concept of a "nuclear winter" that was under study at the time, at the peak of the cold war between the United States and the Soviet Union. Scientists used computer models to run scenarios of partial to full-out nuclear war and calculated the number of firestorms, volume of soot lofted into the stratosphere, and resulting magnitude and duration of the luminosity and temperature drop.

As scientists began adapting the nuclear-winter scenario to the end-Cretaceous impact, they soon discovered that another amplifying mechanism came into play: the searing heat caused by the ejecta as it reentered the atmosphere at full speed all around the globe. Unlike during their launch from ground zero, where the blast had punched a hole in the Earth's gaseous blanket, the rocky missiles were now encountering air resistance and friction on their way back down to Earth.

Calculations geophysicist H. Jay Melosh ran to assess the damage suggested that, all around the globe, the amount of energy dumped into the high atmosphere by the incoming debris would beam down approximately 10 kilowatts per square meter at ground level (more than 1 kilowatt per square foot), raising the temperature to a whopping 250°C (482°F) over a time interval of several dozen minutes. This energy flux can be compared to twelve million (megaton-class) hydrogen bombs, spaced 6 kilometers (4 mi.) apart over the entire globe, firing off simultaneously. The heat pulse would have been sufficient to spontaneously ignite dry leaves and pine needles and set off forest fires around the planet.

According to those early estimates, a large fraction of the Earth's biosphere — plants and animals alike — would have grilled to death and gone up in smoke. But nothing is that simple. First, the calculations themselves are questionable. H. Jay Melosh later refined his model, taking into account the shielding effect that spherules, after delivering their heat and dropping through the atmosphere, would create for other spherules plunging to Earth in their wake. Such self-screening would substantially lower the amplitude and duration of the heat pulse at ground level. On the other hand, fine dust floating in the upper atmosphere might well act as a mirror and reflect more of the spherule heat downward.[17]

Weather patterns also played a role in the heat distribution. Simple models assume an initially clear and transparent sky. But any fog or cloud cover would act as a screen, as constituent water droplets absorb a great deal of incoming heat when they are forced to vaporize. On that fateful end-Cretaceous day, overcast skies would have acted as a protective shield for plants and animals below. Unlucky were the dinosaurs that strolled under blue skies or clear, starry nights.

Is there any evidence in the sediment record for the occurrence and extent of such a global heat pulse? Yes and no: the data is ambiguous. As early as 1985, geochemist Wendy Wolbach published analyses of the K-Pg clay showing a substantial amount of carbon-rich soot — about 1 percent by mass — allegedly caused by the combustion of organic material all around the globe. Scaled to the entire surface of the Earth, the total amount of soot was estimated at 70 cubic kilometers (17 cu. mi.), equivalent to 10 percent of today's biomass going up in smoke. Other scholars identified specific organic molecules in the K-Pg clay, such as retene, found in resinous woods and known to be an indicator of forest fires. On the other hand, critics pointed out that uncharred bits of organic material were found in many continental clay samples, which appeared to contradict the occurrence of global fires. One possible explanation is that, since continental K-Pg clay formed in lakes and marshes, these uncharred items were protected from combustion by water. Not all biomass burned, just a great deal of it.[18]

There is another important piece of evidence that points to global wildfires at the time of impact. The first continental sediments to form right above the clay layer bear little pollen from the rich and diverse assembly of flowering plants that thrived in the Late Cretaceous. Botanists observe that more than 90 percent of plant microfossils after the impact are fern spores. It so happens that ferns are one of the first types of plants to grow after a forest fire, and this "fern spike," as it came to be known, is found all around the globe.[19]

The Big Freeze

Whereas the jury is still out on whether firestorms were global or sporadic in the aftermath of the impact, there is no question that a large amount of soot was lofted into the atmosphere, adding to the opacity caused by the infalling ejecta and dust. As already pointed out, such a darkening of the skies, leading to a protracted nuclear-winter-like event and a shutdown of photosynthesis, was the first killing mechanisms proposed by Luis Alvarez and coauthors when they linked the asteroid impact to the end-Cretaceous mass extinction.

The fine soot measured in the K-Pg clay amounts to at least 15 billion metric tons worldwide, a figure used in computer models to estimate the amount of darkening and cooling that ensued. The simulations showed that the amount of sunlight reaching the surface of the Earth dropped to moonlight levels for approximately one month, remained 99.99 percent blocked for over six months (only 0.01 percent sunlight passing through), and took eighteen months to rebound to 1 percent of its former intensity: the "twilight level" normally observed 200 meters (660 ft.) under water, as well as the minimum amount of light needed by plants and plankton to perform photosynthesis. In other words, photosynthesis would have been completely shut down around

Soot and dust released by the asteroid impact caused obscuration and a severe temperature drop, contributing to the end-Cretaceous mass extinction. Illustration by Willgard Krause via Pixabay.

the globe for a year and a half. By the beginning of the third year, enough soot had settled for the luminosity at ground level to reach that of an overcast winter day, and from then on, sunlight steadily rose back to preimpact levels.[20]

Soot did more than simply darken the skies and interrupt photosynthesis (leading to a temporary collapse of the food chain), which was already lethal enough. It also caused worldwide temperatures to plummet. Computer models suggest that temperatures dropped an average of 15°C (27°F) worldwide for over two years. More precisely, the drop reached 28°C (50°F) on continents and 11°C (20°F) in marine surface waters—the oceans cooling less because of their high thermal inertia. Diffusing downward through the water column, the cold wave still lowered oceanic temperatures by as much as 5°C (9°F) at a depth of 150 meters (500 ft.), according to the models.

Such figures imply that on land subfreezing temperatures prevailed over much of the globe all year long, and lasted several years, except in some coastal areas and a few equatorial regions like Brazil, India, and Indonesia, which were relatively spared, hovering a few degrees above freezing. Thus, already hard-hit by the ejecta heat pulse, the combustion of forests, darkness, the stalling of photosynthesis, and the collapse of the food chain, the biosphere had to deal next with an abrupt ice age, marked by a temperature drop three times greater than any experienced during the recent ice age, over the past 2 million years.

Is there any evidence for the cooling predicted by the models? Fossil plankton and seashells are often used by climatologists to estimate the temperature of the marine waters in which they formed, based on the proportions of various oxygen isotopes in their makeup, which happen to be temperature dependent. When dealing with an event as short as the K-Pg impact, however, it is nearly impossible to find fossil-bearing sediment that accumulated fast enough to offer the kind of fine resolution needed to decipher the climate on the scale of years or even centuries, not to mention the fact that the sea creatures sought for analysis were sparse, massacred by the impact in the first place.

Against all odds, one research team came up with a solution. The Brazos River site in Southern Texas stood at the outlet of a shallow sea—the Western Interior Seaway—during the end-Cretaceous, and it amassed abundant sediment, especially since coastal storms raged in the aftermath of the impact. Instead of relying on rare plankton and seashells, the team collected microscopic fossil archaea, akin to bacteria, that record water temperature by way of the changing chemical makeup of their fatty membranes. In the Brazos mudstone, analysis of these microscopic proxies points to a temperature drop of at least 2°C (3.6°F) in surface waters, with several spikes in the data showing a 7°C (12.6°F) drop, which is in rough agreement with the predictions of computer models.[21]

There is also indirect evidence of cooling. Whereas most plankton species went extinct at the time of impact, surviving cold-water species spread out from the Arctic to colonize all oceans and seas, including the Mediterranean. Fossils of these opportunistic species dominate the first few centimeters (1 in. or so) of sediment above the impact clay, a time interval of roughly 1,000 years. Thus, cool conditions seemed to have dragged on for a considerable length of time.

A Lethal Target

Besides fine dust and soot from forest fires, another climate spoiler was let loose by the impact, and this one had to do with the very nature of the target rock at ground zero. The Yucatán platform is a pile of sediments deposited in shallow water over tens of millions of years, reaching a total thickness of about 3.5 kilometers (2 mi.) and lying on top of crystalline basement rock. Exploratory wells, which were drilled through the platform in the 1950s to look for oil and gas, show that the sediments are mostly carbonates and sulfates. Such carbon and sulfur-rich rocks freed up considerable amounts of carbon dioxide and sulfur dioxide into the atmosphere when struck and vaporized by the impact.

An estimated 100 billion metric tons of sulfur dioxide (SO_2) were released in a mere instant. To put it in perspective, such a figure is equivalent to five thousand times the amount of SO_2 expelled by a large volcanic crisis like the eruption of Pinatubo in the Philippines in 1991, which caused a cooling of the climate of close to 1°C (1.8°F) worldwide for over a year. The reason for the cooling is that sulfurous gases convert to crystals of sulfate high up in the stratosphere: particles that reflect incoming sunlight. After the Chicxulub impact, it took more than two years for the particles to sink below the jet stream and get washed down to the ground, but the cooling effect persisted much longer. According to computer models, the average temperature of the atmosphere dropped by a full 25°C (45°F) three years after the impact, solely as a result of the sulfate load in the stratosphere and independent of any temperature drop caused by the soot of forest fires. Climate experts therefore have two robust mechanisms to explain the cooling recorded in the postimpact sediment.

One should also note that an even greater amount of carbon

dioxide was injected into the atmosphere through the vaporization of carbonates at ground zero: an estimated 1 trillion metric tons. Its greenhouse effect would kick in only after the dust and sulfate veil dropped to the ground at the end of several years, at which time it would help surface temperatures rebound, albeit slowly. Computer models show that the excess 200 ppm of CO_2 added to the atmosphere—twice the amount of CO_2 emitted by humankind since the Industrial Revolution—would have led to worldwide temperatures roughly 2°C (3.6°F) warmer than before the asteroid impact, which is comparable to the increase we will soon experience ourselves with global warming.

These computer models are supported by chemical data, in particular by the isotopic composition of phosphates extracted from fish remains, which records the temperature of seawater. Based on this proxy, it has been suggested that the warming might well have reached 5°C (9°F) for as long as 100,000 years after the impact.

After broiling temperatures, obscuration, deep cooling, and finally global warming, one can add for good measure the collateral damage caused by acid rain. This extra blow came from the sulfurous gases already mentioned, as they converted to sulfuric acid and washed out of the sky, and from additional nitrous oxides produced by the impact's shock wave across the nitrogen-rich atmosphere. Finally, add to this a significant fraction of the carbon dioxide released by the vaporized carbonates, which dissolved into the oceans to form carbonic acid.

Recipes for Survival

With so many lethal mechanisms at play, it seems miraculous that there were any survivors at all. But survivors there were, and species that populate the Earth today are all the offspring of those that made it through the end-Cretaceous mass extinction. Our own ancestor at the time was a small mammal that survived the crisis and ultimately launched the lineage of primates, but we have not yet been able to find a fossil from the period that might fit the bill. In truth, we might never succeed, since well-preserved fos-

sils are rare, and over 95 percent of all species that roamed the
Earth are still unaccounted for. The closest we have found is a genus named *Purgatorius*, whose remains have been found in early
Paleocene strata (past the K-Pg crisis) in Montana and Western
Canada. *Purgatorius* can be pictured as a rat-sized insectivore that
dug small burrows in the ground and climbed trees, and even if
it is not our direct ancestor, it gives us a glimpse of what that ancestor might have looked like (see "Family Portrait" in chapter 3).

Surviving the end-Cretaceous impact was at least in part a
matter of luck. When the asteroid hit, it was not the most "deserv-
ing" species that made it through the crisis on account of some
strength of character or prewired adaptability. It so happened, on
the contrary, that in order to survive, it was a good thing not to be
a big, dominant, sophisticated animal.

One of the first conclusions reached by paleontologists
who tallied the list of surviving species was that all animals that
weighed over 25 kilograms (55 lb.) were systematically exterminated. The great majority of dinosaur species belonged to
this heavyweight category and did not survive. All large marine
reptiles—mosasaurs and plesiosaurs—as well as large flying reptiles of the pterosaur group were also exterminated.

Small mammals boast a larger number of surviving species:
nearly 50 percent made it through the crisis, although their populations certainly dropped to very low figures. Small reptiles,
small amphibians, and fish also fared better than average. What
did size have to do with chances of survival? One can only guess,
but two explanations come to mind: feeding habits and reproduction strategies. One can well imagine that, at the time of impact
and during the days and months that followed, all species were
severely hit, and their populations decimated to various degrees,
but each likely had a certain number of survivors. It was the aftermath of the impact that would determine their fate. In order to
stay alive, individuals had to manage to find food. And for the species to carry on, individuals also had to reproduce.

In both cases, size matters. Large animals need more food than
smaller ones. At a time when the ecosystem had collapsed and
most trees were burned to the ground, an average-sized sauro-

pod dinosaur had to find approximately 25 kilograms (55 lb.) of fresh leaves to eat each day. In contrast, a small rat-size mammal needed only to find 20 or 30 grams (1 oz.) of food per day: one thousand times less.

The other challenge was reproduction. Large-sized species have fewer individuals than small-sized species—a prerequisite for the ecosystem to be in balance, since big animals need wider foraging areas per individual. It follows that if a catastrophe decimates an already-sparse population of large animals, and at the same time further fragments their habitat (in this case through forest fires), the few remaining individuals will find it harder to reconnect and reproduce than a denser (albeit reduced) population of smaller animals would. Another advantage of smaller species compared with larger species is that their gestation period is much shorter on average, and they have a higher number of offspring. Hence, it is faster for them to reestablish a population density that will ensure their survival as a species.

Taking into account this double handicap inflicted by large body size, it is no wonder that no animal with a weight over 25 kilograms (55 lb.) made it through the crisis. Many species happen to be larger today, but they evolved and developed *after* the mass extinction, from survivors that weighed less than the fatal threshold.

A parallel can be drawn with a more recent event: the first phase of the ongoing mass extinction attributed to humans, namely the collapse of the megafauna—mammoths, mastodonts, giant sloths, and the like—at the close of the last ice age (see chapter 4). While the extermination agent is clearly different, one can speculate that the drawbacks of large body size played a similar role: entire species collapsed by failure to reproduce fast enough once the population density had been driven below a critical threshold by overhunting or other causes.

Besides body size, habitat type also played a major role in determining chances of survival. In the continental realm, the survival of freshwater species was much higher (90 percent) than that of land-dwelling animals (12 percent). Such a large difference can be attributed to two factors. One is the protection that freshwater animals obtain from water and water-soaked vegetation against the heat pulse of the impact, caused by fallback ejecta. They needed only dive and stay close to the bottom until the heat wave dissipated. The second factor is that freshwater animals are much less dependent on photosynthetic organisms, which were killed by the low light levels following the impact. Instead, freshwater species rely mostly on detritus food webs—as those who fish with worms, flies, and insects for bait know well—anything that drops accidently into a river. In the aftermath of the impact, there was a giant stock of dead organic matter that was distilled over months or years, blown in by wind or washed in by rain, into the freshwater system.

Similarly, in the oceanic realm, animals that relied on plankton and the photosynthetic food web were the hardest hit, including ammonites, which disappeared entirely at the K-Pg boundary, whereas detritus feeders on the sea bottom showed a better survival rate.

The Lucky Mammals

Besides freshwater species, another group of animals that suffered fewer casualties than average was the mammal class, with half its species making it into the postimpact era. Their small size certainly worked in their favor, as did several other factors. Most mammals adopted nocturnal feeding habits in order to avoid the great dinosaurs they lived side by side with. This led mammals to adapt well to darkness and to the lower temperatures that went with it, an adaptation that included body hair for insulation and a warm-blooded metabolism. When the impact's dust and soot plunged the Earth into a long-lasting frigid night, these traits became invaluable. While surviving dinosaurs were left groping in darkness in search of food, big-eyed mammals had no trouble navigating and feeding among the ruins.

A third advantage had to do with their habitat. A large number of mammal species lived in burrows, as is still the case today. Not only do burrows constitute a hiding place and a defense system against predators but they also offer efficient thermal insulation against temperature swings, protecting their occupants in this particular crisis from both the roasting minutes of the impact's fallback ejecta and from the glacial endless night that followed.

A fourth asset, and not the least, is the feeding regime of mammals. Many of their species are omnivorous, an advantage they owe in part to a diversified set of teeth, including complex, multi-cusped molars. Herbivorous dinosaurs had teeth adapted to specific plants—plants that vanished in the aftermath of the impact. Omnivorous mammals were able to feed on whatever they could muster, be it seeds, roots, or carrion. It was the victory of opportunism over sophistication, the losers surprisingly being the animals that were the most finely tuned to the world . . . a world that no longer existed.

There is irony in the fact that dinosaurs were long believed to be failures of evolution, no longer adapted to the complexity of a world that would inevitably fall into the hands of "superior" beings like us. It now appears the wheel of fortune selected a bunch of cowardly, nocturnal, unsophisticated rodents, who survived by the skin of their teeth.

3 The Road to Recovery

It is important to trace and document at what pace, and in what shape and form, the biosphere recovered in the aftermath of the last mass extinction, 66 million years ago. These recovery patterns might be relevant to the fate of our own ecosystem, if we in turn are to cause a new mass extinction.

The two crises, past and present, might seem to have little in common. The end-Cretaceous mass extinction was primarily based on a brutal roasting phase, triggered by the reentry of glowing ejecta into the atmosphere; a several-year-long obscuration and freezing phase, caused by ejecta dust and soot lofted by wildfires; and a terminal greenhouse warming period, due to the carbon dioxide released from the target rock.

A comparison with our current situation is not entirely irrelevant, however, if we include in today's crisis the threat of nuclear warfare (see chapter 9). A global nuclear exchange would cause wildfires, which would be followed by a nuclear winter triggered by the soot and dust lofted into the atmosphere. Greenhouse warming is already happening, and considering how much energy our civilization has already poured into the climate system, it would resume for some time after the cold snap of a nuclear winter, despite the fact that our industry and agriculture would have collapsed in the process.

Plotting the recovery of the biosphere after a crisis is a difficult task. Even for a relatively recent event, like the end-Cretaceous crisis, one needs a high-fidelity fossil record to follow the tally and evolution of surviving species. In the marine realm, the plankton record indicates that the collapse of the food chain spanned thousands of years, attested by a distinct layer of fossil-poor sediment. This led geologist and oceanographer Kenneth Hsu to pen the expression "Strangelove ocean" to describe the scene, in reference to Stanley Kubrick's cult film, *Dr. Strangelove*, about an accidental nuclear exchange between Russia and the United States. Paleontologist Steven D'Hondt noted that the level of carbonates on the seafloor, a proxy for planktonic marine life, did not return to normal values for approximately 3 million years, but he also made the argument that this low level of planktonic remains could be due to a different ecological network of surviving species in the open waters—one that recycled organic matter before it had a chance to settle to the bottom and make it into the fossil record. As for larger marine animal, they were limited at first to an impoverished assemblage of mostly sponges, brachiopods, corals, and echinoderms—sea urchins and the like, which have an omnivorous, detritus-feeding regime, confirming the selective advantage of scavengers in the aftermath of the impact.[1]

Ocean acidification, the result of the water absorbing excess carbon dioxide out of the atmosphere, also damaged the marine ecosystem, as it threatens to do today. Clay samples from before and after the impact, recovered from a shallow-water cave on the Texas coast, record acidity changes on a very fine scale: the pH (a measure of acidity[2]) of ocean water remained stable over the last 100,000 years of the Cretaceous period (a confirmation, in passing, that volcanic eruptions in India at the time had no noticeable effect on the environment), but there is a jump in acidity of 0.25 pH around the time of impact. It then took about 80,000 years for the ocean's pH to return to normal preimpact levels. It is interesting to note that this 0.25 pH swing in acidity is comparable to the one our oceans will experience over the next century if our civilization does not severely curtail its CO_2 emissions. In other words, if we keep twiddling our thumbs, we will be able to study firsthand a replay of the postimpact acidification event.[3]

Recovery on Land

Species made a faster comeback in brackish and freshwater environments than in the open sea. Acidification was less of a problem there, since limestone banks and other alkaline rock and soil along the waterways acted as a buffer against acid rain. The predominantly detritus-feeding behavior of amphibians, small fish, and reptiles also helped them overcome the crisis. The fossil record in North Dakota shows the reestablishment of numerous turtles, fish, and small crocodilians only a couple of meters (~7 ft.) above the extinction boundary, which means their recovery took less than 50,000 years.

Recovery rates for land animals are more difficult to estimate, because sedimentary environments that capture their fossils in fine detail are much less common. However, there is an excellent outcrop of fossil-rich sediment, east of Colorado Springs, that spans the last 100,000 years before the impact and a full million years past the event. The record shows that raccoon-sized mammals, up to 8 kilograms (18 lb.) in weight, were abundant at the site before the mass extinction, but above the K-Pg boundary, the largest survivors were rat-sized species, weighing less than 1 kilogram (2.2 lb.), which roamed a low-diversity landscape of fern forests. It is not until 20 meters (65 ft.) up the stack of sediment, corresponding to an interval of 100,000 years on this site, that mammal fossils return to precrisis values, both in body size and number of species. By then, plant life had rebounded to sustain larger and diversified species, including palm trees, and according to the type of leaves preserved in the sediment, the air temperature had risen by approximately 5°C (9°F).

Over the next 200,000 years, the fossil record in Colorado shows a further increase in temperature (a couple of degrees) that sustained even more diverse plant life, including walnut-like trees, and a new surge in the diversity and size of mammals. Some were now beaver-size and reached 20 kilograms (44 lb.) in weight. Finally, paleontologists recognize a third jump in temperature and diversity, 700,000 years after the crisis: fossil pea pods show up in the sediment, marking the first occurrence of legumes in North

America. This protein-rich vegetation brought a new boost to the mammal community, with some species now reaching 50 kilograms (110 lb.) in body weight. All in all, it took over half a million years for the ecosystem to fully recover after the end-Cretaceous mass extinction. This serves as an example of how long, in the present era, it would take our biosphere to regain a new equilibrium, if ever humankind brings to completion the sixth mass extinction we might be fostering.[4]

The Roaring Thirties

The era ushered in by the asteroid impact is named the Cenozoic: "new life" in Greek. It is best described as the era of birds and mammals, which replaced amphibians and reptiles as the dominant lifeforms on Earth. Over the course of the next 30 million years—"the roaring thirties," so to speak—the biosphere enjoyed ideal conditions to expand and diversify anew. As often happens after a deadly and destructive war, survivors inherited a land of opportunities. The extermination of numerous species freed up ecological niches on land, in the air, and under water, which birds and mammals readily filled, replacing their deceased reptile rivals.

Mammals in particular seized the opportunity to expand their two major branches: marsupial mammals, where newborns are sheltered and fed in their mother's pouch; and placental mammals, which are more developed and more autonomous upon birth, and which included the first hoofed animals—ancestors of horses, antelopes, oxen, pigs, camels, and the like—and early, small primates that ultimately led to the human species. Fragile primate fossils, however, are rare enough that we might never find the precise forerunners of the human family, but we have a general idea of what our ancestors looked like at the time (see "Family Portrait").

These burgeoning new species were all small to begin with, but they progressively grew larger, in step with their habitat and food sources. Part of this menagerie occupied forests that were claiming back the land at various rates, depending on their climate belt, ushering in the first palm trees, cacti, and many new decid-

Family Portrait

One of the small mammals that appears a mere 100,000 years after the end-Cretaceous impact, and possibly represents one of the earliest primate species, is *Purgatorius*, from North America. It owes its name not to the fact that it swung between life and death at the time of the impact but to the location of its discovery: Purgatory Hill in eastern Montana, infamous for the mosquitoes that harass fossil hunters on the site. Although its fossils are dated slightly after the impact, they obviously do not represent a spontaneously generated new species, but rather one that survived across the extinction boundary. Its preimpact, Late Cretaceous ancestor has not been discovered yet; its discovery would allow paleontologists to assess how much the species changed, if at all, from before the crisis to after.

Based on the few relics discovered, which consist mostly of teeth and jawbone fragments, the small mammal weighed about 40 grams (1.5 oz.) and measured about 15 centimeters (6 in.) in length. Ankle bone fragments support the claim that *Purgatorius* climbed trees, and the shape of its teeth points to an omnivorous diet: partially carnivorous and partially fructivorous, hence its tree-climbing behavior for searching for insects and fruit. The animal probably looked like a small squirrel. If *Purgatorius* is indeed one of the very first "stem" primates, our *Homo* lineage descended from a species very much like it—what a reminder of our humble beginnings.[5]

Purgatorius was one of the earliest primates to evolve after the end-Cretaceous mass extinction. Illustration by Patrick J. Lynch, Yale University.

uous species: bald cypress trees in the wetlands and, in drier climates, laurels, birch trees, and early forms of eucalyptus and other myrtaceous species. Open savannas and prairies also expanded. Such grasslands were already established in the Late Cretaceous, but in the postimpact era, their development and diversification were sparked by the simultaneous rise of grazing hoofed animals: a fine example of coevolution. Vines also appeared on the scene: as fate would have it, the first fossil imprint of a grape leaf was discovered in the Champagne region of France.

Surviving reptiles also underwent a partial recovery, helped by the warm climate that prevailed in the early Cenozoic. Species of crocodilians, tortoises, lizards, and snakes that had passed through the extinction filter on account of their small size expanded anew. One branch in particular literally took off: the bird family. It is now well established that birds descend from a branch of dinosaurs. They had already split off their parental saurian group by the Late Cretaceous, and they weathered the impact crisis much better than their land-dwelling cousins: approximately 50 percent of avian species made it across the K-Pg boundary, a survival rate comparable to that of mammals. This might come as a surprise, considering the noticeable fragility of birds today in the face of deforestation and atmospheric pollution—calamities that also raged in the aftermath of the impact. To solve this apparent paradox, paleontologists speculate that water birds, living in swamps and estuaries, were selectively protected, as were freshwater fish and amphibians, by being independent of the photosynthetic-based collapsing food chain. One can also imagine that birds' flying ability allowed them to migrate from regions most affected by the impact to safer havens.

Be that as it may, the expansion of birds was phenomenal. Because we are self-centered, we like to refer to the Cenozoic as the age of mammals, highlighting our own group, but birds clearly outnumber mammals today—11,000 species versus 6,500—so that, objectively, the Cenozoic should be called the age of birds.

In the shaken-up ecosystem, birds rose to the top of the food chain. Past the crisis, as body size increased once again in the animal world, several bird species reached dinosaur-like proportions. Such was the case with *Gastornis*. Its fossils were first discovered in the southwestern suburbs of Paris in 1855, then in Belgium, Germany, England, and New Mexico. *Gastornis* was 2 meters (6 ft.) tall, flightless, and twice the weight of a human being; it had a large head with a toucan-like powerful, flat beak, and it ran at high speed after its prey. Other equally impressive "terror birds" inhabited Europe, Asia, and North and South America.

In the early Cenozoic, evolution thus favored both dwarves and giants. Animals remained small where food remained scarce and predators were few. But wherever predators started growing in size, it was an evolutionary necessity for prey species to follow suit in order to outrun or fight off the aggressors.

During the first 10 million years of the "roaring thirties," 66 million to 56 million years ago, the biosphere was seen to slowly recover. Soon enough, however, it had to face a new challenge—one of particular interest to our civilization, since it foreshadows the crisis we face today. For obscure reasons and without any notice, a severe pulse of global warming overtook the Earth 56 million years ago, and it had a lasting impact on the biosphere.

Global Warming: A Mysterious Precedent

This remarkable global-warming event was brought into the limelight in 1991 by geologists James Kennett and Lowell Stott, as a result of conducting a routine study of marine sediments near Antarctica. It was known for some time that sea-bottom protozoa (benthic foraminifera) underwent a severe extinction 56 million years ago: half their species vanished over a narrow interval of time. What the new study showed was that the perturbation resulted from a brutal rise in seawater temperature, a fact underscored by the sudden appearance of tropical plankton near Antarctica. The heating pulse became known as the "Paleocene-Eocene Thermal Maximum," or PETM.[6]

The chemical analysis of the sea-bottom sediment and its microfossils revealed that the heating pulse was associated with a massive input of carbon into the atmosphere and oceans. The carbon was richer than usual in its lightest form: carbon-12 (written ^{12}C for short), characteristic of animal and plant tissue, so that it is slightly enriched in fossil bones and shells, coal, and methane.

Today, the concentration of such light carbon in the atmosphere is considered proof of our civilization's contribution to global warming through the burning of fossil fuels.

Certainly, the light carbon spike at the Paleocene-Eocene boundary was not tied to some extinct fuel-burning civilization. There is still much debate with respect to what caused it. Volcanic eruptions were not much favored at first, since their carbon would originate in the Earth's mantle and would not be so enriched in ^{12}C. A preferred idea was that massive amounts of organic carbon were released from fossil deposits, perhaps locked up in ocean-bottom sediment on the continental shelf—methane-rich ices known as clathrates; however, a trigger mechanism was still needed to melt those bottom sediments and release the methane through the water column up into the atmosphere.

Based on the end-Cretaceous example, in which a cosmic impact had devastating effects because it hit a sensitive target, it is tempting to invoke a similar mechanism for the heating pulse of the late Paleocene. An asteroid or comet might have hit a continental shelf rich in methane clathrates, releasing greenhouse gases into the environment. So far, there is no evidence of an impact of the right age on the continental shelf. One crater, however, is dated from the end of the Paleocene, but it is well within the continent: the Marquez Dome in central Texas. It is only 13 kilometers (8 mi.) wide, and so it could not have expelled much carbon into the atmosphere. In addition, marine sediments display a chronology of events that precludes an impact as the main cause of the PETM: in the case of an impact, the outpouring of carbon from the crater would come first, and greenhouse heating would follow, whereas the record shows that the warming of the atmosphere and ocean came first, preceding the massive outpouring of carbon.

From a detailed analysis of the sedimentary record, there is also proof that the crisis occurred in two blows: a small one followed by the large one. The perturbation of the oceanic realm began with an initial temperature rise of 2°C (3.6°F), named the "POE" ("pre-onset excursion"), but the greenhouse carbon was eventually removed from the atmosphere and absorbed into the oceans, and conditions came back to normal. A few millennia later

Different Shades of Atoms

Each type of atom, such as carbon, oxygen, and the like, comes in different varieties, more or less massive, which carry a different number of neutrons in their nucleus: they are called *isotopes* of the element. For instance, carbon has six protons in its nucleus, but it can have six, seven, or eight neutrons. Adding protons and neutrons, these three isotopes of carbon are known as carbon-12 (^{12}C), carbon-13 (^{13}C), and carbon-14 (^{14}C).

Carbon-14 is unstable and decays over time (it can be used as a clock to tell the age of an object that initially contained it). Carbon-12 and carbon-13 are stable but were initially created (by nuclear reactions within stars) in different proportions: approximately one atom of ^{13}C for every one hundred atoms of ^{12}C.

This initial ratio can change during physical and biological processes, a behavior that carries precious information. Living organisms that pump carbon out of the environment to build their cells preferentially use ^{12}C, which is lighter and requires less energy to include than ^{13}C. As a result, animal and plant fossils display a greater $^{12}C/^{13}C$ ratio than nonliving matter. Today, this is one way to tell that the carbon dioxide building up in the atmosphere, which is enriched in ^{12}C, comes from organic matter—civilization's burning of fossil fuels in particular—rather than from volcanoes or other abiotic sources.

On the other hand, oxygen isotopes, which also exist in three forms (^{16}O, ^{17}O, and ^{18}O), provide information about the temperature of seawater. When seawater (H_2O) starts to evaporate, the molecules containing the lighter form of oxygen (^{16}O) evaporate preferentially at first. But if the water temperature continues to rise, more heavy oxygen (^{18}O) enters the evaporation process; in the remaining liquid water, the ratio of heavy oxygen drops accordingly. Plankton and other lifeforms that absorb oxygen from such warm water to build their mineral shells therefore contain less ^{18}O and, when they die, carry this testimony to the ocean floor. The result is that when fossil-rich sediments are cored and analyzed, the oxygen ratio measured in the fossils betrays the seawater temperature at the time. With the help of these chemical proxies, scientists have established that the oceans, and the environment at large, heated up at the end of the Paleocene (oxygen isotope testimony) and that the carbonic gases that were released as a result were organic in nature (carbon isotope testimony).

came the PETM proper, linked to a much larger release of carbon that raised the temperature this time by 4°C (7.2°F).

For both pulses, large volcanic eruptions might be the trigger. The PETM warming event happens during the spreading of a volcanic rift zone between Greenland and Europe, opening up the North Atlantic basin over a time frame of several millions of years. Although the eruption rate of this volcanic rift is believed, on its own, to be insufficient to account for so much carbon released in so little time, one proposed mechanism is that carbon-rich rock of the Earth's mantle, 160 kilometers (100 mi.) or so below the surface, fueled the ascending magma and released extra carbon into the atmosphere. Or else, eruptions would have hit carbon reservoirs at the surface, such as methane clathrates, by chance. Food for thought: the jury is still out.[7]

Global Warming and Evolution

Identifying the initial cause of the Paleocene-Eocene Thermal Maximum is a work in progress. Beyond this mystery, what is particularly relevant to our civilization is how the warming affected the biosphere and how it recovered. For most observers, we are indeed fostering a heating pulse on Earth that will soon reach PETM proportions: in other words, today's runaway global warming might well become the most severe hyperthermal crisis to hit the biosphere in the last 56 million years.

Global warming today might easily match, or even surpass, the level it reached during the PETM. Our civilization pours approximately 10 billion metric tons of carbon per year into the atmosphere, whereas during the PETM, the annual input is thought to have been ten times less. The main difference, so far, is that the PETM kept up its carbon output over tens of thousands of years, whereas our civilization is only a few centuries into its industrial and agricultural carbon emissions, with an output of 600 billion metric tons of carbon so far, versus 10 trillion metric tons for the entire PETM. If we keep up our present rate, however, we will reach PETM global figures in less than 1,000 years.

The situation is all the more critical today because an unexpected acceleration can occur at any time, with global temperatures shooting even higher, through the triggering of some feedback loop, such as the breakdown of methane-rich clathrates on the ocean floor—a mechanism suspected to have occurred during the PETM. The Intergovernmental Panel on Climate Change (IPCC) points to a similar threat today: the permafrost locked up in the tundra and bogs at high latitudes contains some 1.7 trillion metric tons of carbon—twice the amount of carbon currently in the atmosphere. We have no idea at what point the ongoing temperature rise might release the carbon trapped in such reservoirs, but when it does, such a release could occur fast and catastrophically. These are the kind of feedback loops that threaten our world today, in which warming can beget more warming.[8]

The PETM teaches us another lesson: it is a prime example of how the biosphere reacts to a pulse of global warming. One observation is that extinctions were moderate at the time—there was no mass extinction on the scale of those we have discussed so far. Many species possibly had the ability to move to higher latitudes or altitudes to find the right temperature bracket to survive, while others were able to evolve on site, over several generations, to adapt to the changing environment. We should beware that the situation is different today: such mobility is made difficult by habitat loss and fragmentation, as a result of urbanization, agriculture, and deforestation, so that in today's world, many species might be trapped on site and driven to extinction by any significant temperature rise.

This said, short of a mass extinction, the response of the biosphere to the PETM was dramatic nonetheless. An exceptional fossil site in Wyoming, the Bighorn Basin, tells that story. Less than 10,000 years after the onset of the heat pulse, mammals evolved in a flash to produce new versions better adapted to higher temperatures: hoofed ungulates with an uneven number of toes (Perissodactyla), including the ancestors of rhinoceroses (three toes) and horses (one toe); the first American even-toed ungulates (Artiodactyla), which would yield the future families of Bovidae (buffaloes, oxen, sheep, and goats), Suidae (pigs and hogs), and Cervidae (deer and moose); novel versions of primates; and hyena-like carnivores. All these new prototypes were small in size, in response to higher temperatures, since a smaller size guarantees a more efficient heat loss.[9]

All these new species did not pop out of a magician's hat through some sort of spontaneous generation: they expanded out of restricted but prolific sites of evolution, where small populations were inbreeding and churning out new species, some-

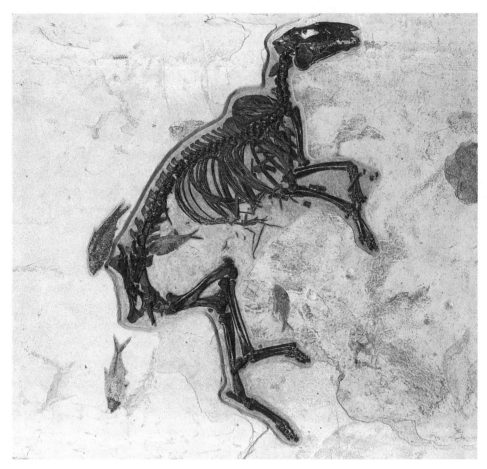

Protorohippus was an early member of the horse family, 30 centimeters (12 in.) long, that evolved after the Paleocene-Eocene faunal turnover. Photo courtesy of National Park Service.

where in Mongolia or Southern China. It still remains a mystery why global warming dislodged them from their hiding places, but as a result, they spread across Asia, Europe, and North America, and they radically changed the balance of the ecosystem and the course of evolution on Earth.

The exceptional heating pulse of the late Paleocene and the resulting change of cast in both animal and plant life were completed in less than 200,000 years. The situation then stabilized, ushering in the Eocene period, 56 million to 34 million years ago, when tropical and subtropical forests took over the land. This Garden of Eden lasted over 20 million years. But all good things come to an end. After the warm climate that prevailed during the first half of the "age of mammals," the second half was shaken up by a series of cooling steps that ultimately triggered an ice age.

Cooling the Earth

If we wish to better manage our planet, we must first understand how it functions: how it heats up, and also how it cools down. The last few million years are particularly relevant in this respect, since they are marked by a spectacular sequence of cooling steps, some rather brutal, others more progressive, with multiple causes.

As early as the nineteenth century, paleontologists noticed in the fossil record that the flora and fauna of the warm Eocene period began shifting at some point toward species better adapted to more temperate climes. They defined this transition as the start of a new geological period, roughly 34 million years ago, which they named the Oligocene.

Although it often appears pinned down to one date, the transition is progressive and complex, contrary to the brutal end-Cretaceous boundary. But since the latter had made such an impression on researchers and science reporters, many wished to view the Eocene-Oligocene transition as an abrupt crisis as well. Dating sediments and fossils around the globe came with its share of inaccuracies, making it possible to cluster into brief intervals of time events that might actually be millions of years apart. Geologists were all the more inclined to espouse a late Eocene catastrophe when, during the 1980s, they discovered traces of iridium and shocked quartz in the sediment spanning the boundary, mimicking the impact signature of the end-Cretaceous. A careful analysis ended up revealing two distinct spherule beds several centimeters (a few inches) apart—as if two asteroids had hit the Earth less than 20,000 years from each other.

It was not long before the impact sites were discovered. The older spherule bed was traced to Popigai crater in Siberia, which was 90 kilometers (56 mi.) wide, and the younger one to a buried structure under Chesapeake Bay, south of Washington, DC, estimated to have been 40 kilometers (25 mi.) wide at the time of impact, although it slumped to form a bowl twice that size. Both craters are dated at 34 million years. The fact that the impacts correspond to a cooling trend on Earth might be more than a mere coincidence. These impacts might be part of a salvo of an even larger number of impacts at the time, the others much smaller and beyond recognition today; that is, if we are to believe the increase in the amount of helium-3, a proxy for cosmic dust, that spans a long interval of 2 million years in the sediment at the time.

Besides the fact that there was a cooling trend across the protracted Eocene-Oligocene boundary, the evolutionary shift in plant and animal species came with its share of extinction pulses, albeit short of a true mass extinction. The Popigai and Chesapeake Bay impacts are not directly associated with any particular extinction pulse: at best, they coincide with some noted shifts in the proportions of plankton species, and hence of sea water temperature, but that's about all. This goes to show that a significantly larger impact than Popigai (at the K-Pg boundary, Chicxulub was arguably ten times more powerful) is needed in order to decimate the biosphere.

Volcanic eruptions do not appear to have played a significant role either. It was once thought that large volumes of basalt extruded in Ethiopia were synchronous with the Eocene-Oligocene transition, but they are now found to be much younger eruptions, peaking around 30 million years ago.

Whatever caused it, the Eocene-Oligocene turnover of species occurred in several pulses, spread out over 7 million years: terrestrial mammals underwent a first crisis around 40 million years ago, followed by marine animals 37 million years ago, and a second mammal crisis 33 million years ago.

The first installment targeted not only mammals but reptiles and amphibians as well. Tropical forests and swamps were affected the most when the temperature plummeted by a full 10°C (18°F): a spectacular shift, since in today's world, the temperature drop from a warm interglacial to a glacial period does not exceed 5°C (9°F). In North Dakota, for example, according to fossil leaves, dense tropical forests were replaced by scattered deciduous trees; this points not only to global cooling but to more contrasted seasons as well, with longer dry spells. Salamanders, crocodilians, and turtles declined abruptly. Tree-climbing mammals were also hit: in North America, close to half the species disappeared. In Europe, the expansion of monkey-like primates came to a halt: the trend reversed, and their numbers started dwindling. In the changing forests, browsers accustomed to succulent tropical plants were replaced by species better adapted to tougher leaves.

In the marine realm, the faunal turnover seems to kick in a couple million years later, around 37 million years ago, when a radical change took place in the plankton world: tropical species were replaced by their high-latitude cousins, which were better adapted to cold water. The cold spell was particularly lethal to gastropods and bivalves (80 percent species loss), which makes this turnover the most notable crisis in the marine world since the end-Cretaceous mass extinction 66 million years ago.

The extinction rate then dropped back to normal for the next 3 million years—during which time the Popigai and Chesapeake impacts took place with little apparent damage, as mentioned—but the truce was broken by a further drop in temperature, starting 33 million years ago. Marine life was decimated anew; on land, climate proxies indicate a worldwide drop of 13°C (23°F), underscored by a deepening contrast between summer and winter temperatures.

The new cold spell did not seem to affect North America too much, although massive browsers, such as the rhino-like brontotheres, did go extinct at the time. Perhaps the continent's relative isolation prevented the intrusion of foreign animals that could have challenged and driven local species to extinction. In Europe, it was another story. The rate of species extinction and turnover is so marked that the transition was nicknamed the *Grande Coupure* ("the Big Break") by Swiss paleontologist Hans Stehlin in 1910. Numerous mammals were affected, including—again—the tree-climbing primates, which were, this time around, completely eradicated from Europe, never to return.

The turnover is all the more striking since, among the mammal group, species that underwent extinction were replaced by new species that were clearly Asian in character, including sophisticated rodents, ancestors of the future hare and rabbit family, and early prototypes of the rhinoceros group. All were well adapted to cold weather and to feeding on tough leaves and shrubs. Perhaps these species originated in the high plateaus of the Himalayan range, where they were accustomed to coldness and temperature swings typical of a mountain climate, in contrast to "old school" European mammals that could not cope with the cooling.

Be that as it may, the Eocene-Oligocene crisis deeply transformed the biosphere in a matter of a few million years, as the world cooled down, step by step, from a tropical Garden of Eden to a much harsher environment. This might seem irrelevant to us today, when humanity faces the opposite trend, struggling to understand the mechanism of global warming. But if we are to fully understand our planet and its changing climate, both warming and cooling trends are important. Some computer models even show that a warming spell can trigger a much more damaging cooling spell in its wake, if the Earth system overreacts.

The Role of Antarctica

The cooling trend that began at the end of the Eocene was a long-lasting one. It certainly brought into play slow and lasting mechanisms: one appears to be the slow motion of tectonic plates and continental drift, affecting ocean currents to the point of upsetting the climate.

Take the opening or closing of straits between continents. Given the slow pace of tectonic plates—moving only a few centimeters (about an inch or two) a year—opening or closing straits seems to be an excruciatingly slow process, but on geological timescales, it can actually be quite rapid. The Strait of Gibraltar, connecting the Atlantic Ocean and the Mediterranean Sea, is only 15 kilometers (9 mi.) wide. At the pace at which plates converge or separate, the closing of the Gibraltar gateway could take less than 500,000 years. Its effect on climate might even be shorter, since the strait need not be completely closed for the oceanic connection to be severely curtailed and the global climate affected.

The end-Eocene climate drop might have been caused not by the closing but by the opening of straits between South Africa, Australia, and Antarctica—continents that were beginning to separate at the time. Until then, Antarctica was bordered by relatively warm currents that traveled down the Atlantic, Pacific, and Indian Oceans and curled around the continent. When Antarctica separated from Australia and Africa, the oceanic circulation pattern was deeply affected: a cold current began swirling around Antarctica, isolating it from the rest of the system and transforming it into a cold trap. Any humid air reaching the polar continent froze, precipitated as snow, and compacted into an ever-expanding ice cap. That the cooling trend at the time occurred in several steps might also reflect the fact that Antarctica first separated from Australia, then from Africa. Be that as it may, Antarctica's ice cap first grew on the eastern half of the continent, then on the western half.

Changes in oceanic circulation, evaporation, and precipitation come with a number of side effects. If evaporated seawater precipitates as snow onto an ice cap, becoming sequestered without returning to the ocean, the sea level drops. And as the sea level drops, more continental shelf is exposed. Because the exposed sea bottom is lighter in color than the deep blue water that used to cover it, the net effect is a brightening of continental margins, reflecting more sunlight out to space, which further drops global temperatures. This is another example of a positive feedback loop, or "snowball effect."

A drop in sea level also has a direct impact on sea life. If the submerged continental shelf narrows, there is less space for fish and shellfish, leading to increased competition between species and more extinctions. This might explain, in part, why the cooling trend of the late Eocene is also a time when gastropods and bivalves significantly declined.

Another contribution to the snowball effect is less obvious. When, during a sea-level drop, a new swath of continental shelf is exposed to air, its rocks become altered by atmospheric gases, in particular carbon dioxide that becomes locked up in the altered minerals. Hence, carbon dioxide is pumped out of the atmosphere, weakening the greenhouse effect and further bringing temperatures down. Not to mention the colder the seawater becomes, the more carbon dioxide it can absorb, likewise pumping it out of the atmosphere.

The end-Eocene example serves to show how an initial cooling step can escalate into a full-fledged "ice box" situation, through a complex set of chain reactions. Our civilization should pay attention to these past episodes and strive to understand how they operate, as we are confronted today with similar chain reactions, albeit in the opposite direction—global warming.

Homo Takes Center Stage

Once past the onset of global cooling during the Eocene-Oligocene transition and the extinctions that ensued, the biosphere settled into a new routine. Temperatures kept declining, but animal and plant communities adapted and remained fairly stable throughout the cool Oligocene, Miocene, and Pliocene periods, without any major upheavals, from roughly 34 million years ago until the end of the Pliocene, 2.6 million years ago.

At least, that's the big picture. On a smaller scale, interesting events took place in specific regions and among certain groups of species. One such region was Eastern Africa, where a mantle plume—an anomalous batch of rising hot rock—began to uplift and stretch the crust, around 30 million years ago. Faults propagated northward, opening the rift zone of the Red Sea that ultimately separated Arabia from Africa. Faults also propagated southward across East Africa, diverging into two branches. The eastern branch tore through Ethiopia, Kenya, and Tanzania; the western branch spread across Rwanda and Burundi and followed the border between Tanzania and the Democratic Republic of the Congo before both branches converged southward to merge in Mozambique.

The establishment of the East African Rift system upset both the regional topography and climate. Within each branch, the

The East African Rift was the cradle of australopithecine evolution (*inset, Australopithecus afarensis*), which led to the human genus. Photo by Roma Neus, licensed under CC BY 3.0; inset photo by Pbuergler, licensed under CC BY 3.0.

Earth's crust stretched and collapsed, stepping down toward a central valley, while the flanking shoulders of the rift underwent an opposite, compensatory uplift. As they slowly rose, over hundreds of thousands of years, the escarpments altered the wind regime and rainfall, creating dry zones on their lee side. Tropical forests, which until then had been a haven for tree-climbing primates, began to decline, progressively being replaced by open grasslands.

As early as the 1970s, geologists began to suspect that the evolution of the human family in the East African Rift, at a time when its uplift was changing the regional climate, might have been more than just a coincidence. As a student in plate tectonics at the time, I ran across an article that first hinted to such a link,[10] and it prompted me to spread the concept in a lecture I delivered in Paris in 1979, at a symposium on the emergence and evolution of intelligence. One of the senior paleontologists attending the conference, Yves Coppens, expressed skepticism at first but endorsed the idea in 1982, coining for it the clever name of "East Side Story." In the meantime, I had publicized my initial thoughts in my first book, which painted the following picture (translated from the French):

In order for the human branch to develop, there had to be an upset, a change of scenery that put tree-climbing primates in jeopardy and steered evolution toward creatures better adapted to the changing conditions: the *Homo* genus.

It so happened that such a radical change of scenery did take place in Africa over the past several million years: the creation of the Great Rift Valley. Accompanied by earthquakes and volcanic eruptions, the rifting apart of the African plate triggered a complete overhaul of landscape, climate, and vegetation. Tectonic motions caused the central valley to subside and the flanking rift shoulders to lift up, casting their eastern slopes into rain shadow.

Such an overhaul did not happen overnight. The change of scenery was progressive, giving animals enough time to adapt. The great tropical forests, which used to stretch out in all directions, were doomed. Changes in topography and climate led to their decline, and in their stead, patches of grassland began to spread. This was an unexpected turn of events for primates, who were faced with shrinking forests: soon there was not enough room for them in the treetops, and food became scarce, forcing them down into the savanna. Perhaps because the tall grasses blocked their view, our distant ancestors were forced to stand upright to better spot predators and to look for food.... Stimulated by all these new developments, the brain of bipedal species grew larger, at an extraordinary pace. Around 3 million years ago, the human genus was well on its way, creating its first tools.[11]

This is a crude scenario, but recent research appears to support the influence of tectonic uplift and climate change as an important, and perhaps a decisive, factor in the evolution of the hominin branch. The first unequivocal bipedal specimen in our ancestry is the australopithecine Lucy (*Australopithecus afarensis*), discovered in 1974 and dated to about 3.2 million years ago. But the story gets even more interesting at a later date.

In 2002, a team of paleontologists suggested that rather than the overall change of climate affecting the Rift during this period, it was its periodic oscillations—the repeated pulses of change— that drove human evolution many steps further. By analyzing lacustrine sediments and their fossils in the Turkana Basin, on the border of Ethiopia and Kenya, the team noted an important climatic variation, 2.8 million to 2.6 million years ago: a particular pulse in the decline of tropical forests, replaced by more scattered and varied woodlands. This time interval happens to coincide with the rise of *Paranthropus*, a branch of hominins that evolved from australopithecines and clearly shows a bipedal morphology, although it did not end up giving rise to our *Homo* genus and is rather considered to be a sister group.[12]

In March 2015, another team announced the discovery, in the Afar region of Ethiopia, of a fossil jaw, dated 2.8 million years ago, that appears to represent a transitional species between Australopithecines and *Homo*: the fossil is associated with bone remains of other mammals that point clearly to a more open and drier environment. In the environmental upheaval that followed, 2.4 million to 2.2 million years ago, which was marked by a clear expansion of savannas, *Homo habilis*—the first true representative of our genus—appeared on the scene, 2.33 million years ago. Its brain size reached 600 cubic centimeters, whereas the australopithecine Lucy, 1 million years prior, had a cranial capacity around 400 cubic centimeters, and the species is credited with the first stone tools, discovered in sediments of the same age: choppers and other roughly carved chunks of quartz and basalt used to butcher prey, crush bones, and skin furs. During the next climate swing, 2 million to 1.8 million years ago, when the savanna became widespread (as attested by the many fossils of grazing mammals), *Homo erectus* was next to enter center stage. Our close cousin now boasted a cranial capacity between 750 and 1,000 cubic centimeters, was long-legged (which helped to cover long distances), and mastered an advanced lithic industry, boasting elegant and efficient bifacial tools (hand axes, carved on both sides).[13]

In view of these discoveries, the stepwise evolution of

hominins appears to have been synchronous with the climatic pulses that affected the East African Rift. Such a pattern suggests a causal relationship—at least in part, since many other factors certainly converged to drive the evolution of the human genus.

The Ice Age

While the destiny of the human genus was playing out in East Africa, the Northern Hemisphere was also experiencing its own climate oscillations on a much larger scale, beginning 2.58 million years ago. The changes were mostly felt in Eurasia and North America: it was a new chapter in the history of the Earth, defined as the Pleistocene epoch, more widely known as the Ice Age. The name is somewhat misleading: although global temperatures continued to drop, a closer look shows that the climate alternated between cold glacial periods, or "stages," and slightly warmer interglacial ones.

The beginning of this new drive toward lower temperatures is fairly well constrained. Whereas the earlier long-term cooling of the Earth that started at the Eocene-Oligocene transition, somewhere between 40 million and 35 million years ago, can be attributed to continental drift separating Antarctica from Africa (placing Antarctica over the South Pole and transforming it into a giant icehouse), the new drop in temperature, which began approximately 3 million years ago, is blamed on continental convergence, rather than divergence.

Up until then, North and South America were separated by a loose string of islands that allowed strong currents to flow eastward from the Pacific into the Atlantic Ocean, strengthening a global equatorial trend and limiting the extent of north–south currents. These Central American straits began to close, however, through the motion of crustal blocks rising along tectonic faults, filling gaps between the islands and building the Isthmus of Panama that ultimately joined North and South America.

The exact timing of the buildup is still a matter of debate, but the final outcome was a global reorientation of ocean currents: their equatorial flow now interrupted, they began to swirl instead from equator to pole, most notably in the Atlantic Ocean, where warm currents and air masses circulated toward the Arctic. When this moist air reached high latitudes, snow began precipitating over Greenland and neighboring islands. Another contributing factor was the slow shifting of Greenland into an Arctic position closer to the pole, ready to collect the incoming snowfall.[14]

After Antarctica, the Earth thus inherited a second icehouse, with Arctic ices reflecting an extra share of sunlight back to space. The planet experienced a new temperature drop of 2°C (3.6°F) in the North Atlantic.

Once Greenland and the Canadian High Arctic developed an ice cover and sea ice spread over the Arctic Ocean, a climate equilibrium was reached, in which the climate oscillated between two states: modest phases of Arctic glaciation, known as interglacial stages, and more pronounced buildups of ice, spreading farther south, termed glacial stages.

Climate oscillations are nothing new in the history of the planet, but this one is particularly striking. While it is commonly accepted that the change in ocean currents 2.5 million years ago amplified the mechanism, its remarkable periodicity is caused by cyclic variations in the motion and attitude of the Earth as it orbits the Sun. If we wish to pinpoint the root of the problem even further, it is our sister planets that are to blame. The Earth would follow a fairly stable and straightforward orbit if it weren't for its distracting neighbors. Venus, Mars, and even the distant but massive planets Jupiter and Saturn play tug-of-war with the Earth, exerting a slight gravitational pull that causes our world to wobble and steer off course.

Danish geologist Jens Esmark was first to suggest such interferences, as early as 1824, and French mathematician Joseph Adhémar succeeded in describing them mathematically in 1842, but it was Serbian geophysicist Milutin Milankovitch who went down in history by publishing a detailed account of these astronomical cycles in 1924, known today as Milankovitch cycles. It is worth recalling them briefly since they form the basis of the long-term climate variations that the Earth experiences today.

Half a dozen astronomical cycles are superposed, but three

stand out. The best-known cycle is axial precession. It occurs because the Earth's rotation axis is tilted, rather than straight up, relative to the plane of Earth's orbit around the Sun: currently, an angular tilt of 23 degrees and 27 minutes of arc (23°27′). Today, our rotation axis happens to point toward a star in the Little Dipper, known as the North Star, or Polaris. This axial tilt is what causes seasons on Earth: as we orbit around the Sun, the fact that we invariably point in the same direction causes the tilted Northern Hemisphere to face the Sun part of the time, with the extra insolation causing summer (whereas the Southern Hemisphere turns its back to the Sun, experiencing winter). Six months later, on the opposite side of our solar orbit, roles are reversed: it is now the Northern Hemisphere that turns its back to our star, experiencing winter, and the Southern Hemisphere that points toward it.

Such would be the normal course of the seasons, but the rotation is altered by a combination of two factors: the Earth is not a perfect sphere, which creates a wobble, and the gravitational attraction of other planets contributes to this wobble. The axis of the spinning Earth slowly swirls around like the axis of a toy gyroscope, drawing a cone in space and returning to its initial position, again pointing toward the North Star, over a period of 26,000 years. As such, this shift in the pointing does not affect the insolation curve and the march of seasons: it progressively shifts the place along the Earth's orbit—and thus the day of the year—when a given season begins and ends: the dates of solstices and equinoxes.

Things become interesting when we add a second cycle: the fact that Earth's orbit is not a perfect circle but a slightly stretched ellipse: we are 1 percent farther from the Sun—about 1.5 million kilometers (930,000 mi.)—in July than in January, which means that, today, summer in the Northern Hemisphere is slightly cooler than average (and winter in the Southern Hemisphere slightly warmer). However, since the timing of seasons shifts along our orbit, roles will be reversed in roughly 13,000 years (half a cycle of axial precession), with Northern summer getting in turn that 1 percent of extra warming.

The difference is slight indeed: a distance 1 percent closer or farther from the Sun only equates to a variation of 2 percent in insolation. The figures, however, can get more extreme, and this is where the second Milankovitch cycle comes in: the ellipticity of our orbit varies periodically. From a perfect circle (zero eccentricity), it expands past the present state (1 percent variation between closest and farthest distant to the Sun) to reach a maximum asymmetry of 6 percent, before returning again to zero, over a period of roughly 100,000 years. During maximal eccentricity, the insolation reaching Earth varies by 12 percent over a year, which is much more than at present.

Finally, the third cycle that comes into play is the inclination of the Earth's rotation axis (obliquity), which, like a wobbling gyroscope, does not remain fixed at 23°27', but nods up and down, reaching 25° at times before straightening up to 20°, over a period of 41,000 years.

Combine all three cycles—axial precession, eccentricity, and obliquity—and insolation can vary by as much as 22 percent over thousands of years at the level of the Arctic Circle. Just as important is the month of the year when minimum or maximum insolation is reached. This is how the Earth enters a glacial or an interglacial stage. When the lowest point of the insolation cycle corresponds to summertime in the Arctic, the snow falling during the winter does not melt in the summer, and the ice cap grows from year to year, starting a glacial stage by reflecting more sunlight back to space. In Europe, the ice cap can extend southward through Scandinavia and the British Isles, down to France and Germany. After several tens of thousands of years, eccentricity, obliquity, and axial precession will add up to instead trigger the opposite effect: from minimal, insolation becomes maximal over the Arctic during summertime and the ice cap starts receding, year after year. As the ice cover shrinks and more sunlight is absorbed by the uncovered, darker land, the climate warms, and the Earth enters an interglacial stage, as is the case today.

One might wonder why the Ice Age, with its alternating cooler and warmer stages, only concerns the last 2.5 million years, since Milankovitch cycles have been active throughout the history of the Earth. The reason is that their effect is subtle, with little consequence, except in a particularly favorable lineup of circumstances. It was the closing of the straits between the Americas, driving ocean currents and moist air toward the North Pole, as well as the proximity to the North Pole of land masses where ice could build up—most notably Greenland and Scandinavia—that tipped the balance.

Over the past million years or so, glacial and interglacial stages have settled into a periodicity of about 100,000 years, matching the eccentricity variations of the Earth's orbit, which seems to have become the most influential of the three Milankovitch cycles. Another interesting detail is that within a cycle, the lengths of the glacial and interglacial stages are strongly asymmetric, with glacial stages typically lasting 80,000 years or so and interglacial ones less than 20,000 years. The pace of the switchover is also different. The onset of a glacial stage is relatively progressive, whereas the transition to the warmer interglacial is rather abrupt.

Hence, the last cycle began progressively, 120,000 years ago, introducing a long glacial stage during which, incidentally, the last branch of the human genus, *Homo sapiens*, left the East African Rift to spread across Europe, Asia, and even the Americas, hunting as far north as the edge of the ice caps. It was only 12,000 years ago that the long cold spell suddenly came to an end and that, in the much more favorable climate of the present interglacial, humankind developed agriculture, raised livestock, founded the first cities, and established modern civilization.

4 Humankind and the Extinction of the Megafauna

The onset of the Ice Age, roughly 2.5 million years ago, mostly affected the Arctic. Meanwhile, the human genus kept evolving in Africa, diverging into multiple branches of competing species that progressively mastered stone toolmaking. In particular, the fossil bones of *Homo habilis* are associated with stone tools carved on a single face, their first occurrence dated at 2.5 million years ago. Next on the scene appeared *Homo ergaster*, approximately 2 million years ago, a species that stood completely upright and had a more voluminous braincase, followed soon thereafter by *Homo erectus*, a descendant or cousin that migrated to Asia. These species are credited with having more complex societies, better toolmaking skills—stones carved on both faces—and perhaps a rudimentary form of language and organized hunting strategies.

Little by little, wave after wave, these hominins settled Europe, and each new set of slightly different fossils inherited a new species name and emplacement on our phylogenic tree: *Homo antecessor* in Spain, believed to be the oldest European on record, and *Homo heidelbergensis* in Germany (600,000 years ago), along with its more recent French representative (450,000 years ago) discovered at Tautavel, at the foot of the Pyrenees mountains. Paleontologists broadly agree that all these species belong to an evolutionary branch that ultimately led to *Homo neanderthalensis*, commonly known as Neanderthals.

This genus was a very promising prototype of modern human but not the final version that fostered human civilization. Out of the *Homo ergaster/erectus* genetic pool still in Africa emerged a new offshoot, around 300,000 years ago: a taller, slimmer species, with an even larger braincase than its cousins. Named *Homo sapiens*, this new species also migrated in several waves out of Africa, starting around 100,000 years ago. It moved first through the Middle East, toward Asia and Australia, and then into Europe, around 40,000 years ago, where it shared the land with its precursor, *Homo neanderthalensis*. Both species cohabited for a few thousand years, and some interbreeding even occurred between them: modern humans, especially in Europe, have up to 3 percent of Neanderthal genes in their DNA. *Homo neanderthalensis* then disappeared from the scene, for reasons that I will address in chapter 8, while *Homo sapiens* kept developing toolmaking and organized hunting, stalking the large mammals—elk, bison, and mammoths—that roamed the cold steppes on the fringe of the ice caps.

The Megafauna

Meanwhile, a rich and diverse fauna of large animals had flourished during the Ice Age, adapting to the cold climate and to a vegetation cover consisting mostly of hardy bushes, grass, and moss. These animals did not come out of nowhere. Each species had its own history. Close cousins of elephants, mammoths originated in Africa and reached Eurasia 3 million years ago, weathering several glacial and interglacial stages and branching out into many different species, two of which prospered in Europe, starting approximately 600,000 years ago: steppe mammoths and woolly mammoths.

Other life histories are equally interesting. The Irish elk, *Megaloceros giganteus*, over 2 meters (6 ft.) tall at the shoulders and carrying the largest antlers of any known deer, likely originated in the Middle East before spreading into Eurasia; the steppe bison, *Bison priscus*, probably originated in South Asia before spreading to Europe 900,000 years ago, where it is well represented in cave art, and then moving on to North America; and the woolly rhinoceros, *Coelodonta antiquitatis*—boasting a long, flat horn to plow the snow while searching for food— probably originated on the Tibetan Plateau before reaching Europe 200,000 years ago.

Many species, like the woolly mammoth and the woolly rhinoceros, were adapted to cold weather and prospered in Europe during glacial stages, while others, like leopards and hippopotamuses, took advantage of the warmer interglacial stages to migrate north and mingle with their woolly neighbors before retreating back south when the cold weather returned.

What most of these animals had in common was their large size, so that they are collectively known as the Ice Age megafauna. To qualify for the name, the adults of a species had to weigh at least 50 kilograms (110 lb.), but many weighed more than a metric ton (1,000 kg). Mammoths even weighed 5–6 metric tons, as do African elephants today.

These megafaunal species left easy-to-find fossils because their bones are large and recent. Some individuals have even been preserved whole, including flesh, skin, and fur, frozen for posterity in the Arctic permafrost; pickled in brine, as happened to a woolly rhinoceros now on exhibit in the Cracow Museum; or retrieved from deadly sink holes, most famously at La Brea tar pits in Los Angeles. Moreover, our *Homo sapiens* ancestors painted and carved remarkable portraits of mammoths, bison, aurochs, and many other species on the walls of caves in Spain and France, most famously at Lascaux, Chauvet–Pont d'Arc, and Cosquer.

Ice Age megafaunal species display some notable differences with large animals living today, but the major difference is that they are now dead. Confronted with megafaunal skeletons, nineteenth century naturalists were at odds to explain the sudden and recent disappearance of these species, and many invoked the biblical Flood to solve the issue, setting off a debate that pitted religious explanations against the revolutionary concepts of evolution and mass extinction. Today, the demise of the megafauna is at the center of just as controversial a debate. Although it has become clear (to everyone except die-hard fundamentalists) that

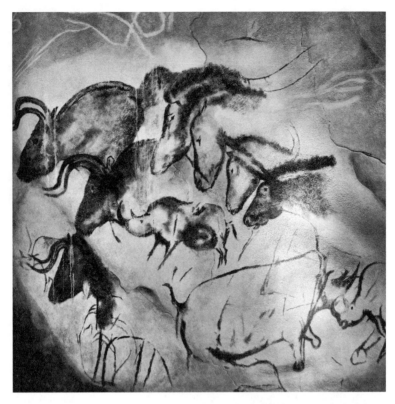

Prehistoric art from the Chauvet–Pont d'Arc cave (Ardèche, France), dated approximately 35,000 years before present, displays aurochs (ancestors of modern cattle), horses, and rhinoceroses, all part of the European megafauna. Photo by Thomas T., licensed under CC BY-SA 2.0.

no god or flood is to blame, the question has shifted to the responsibility of Darwinian natural selection versus the impact of human civilization.

Humankind Accused of Genocide

While the precise timing differs between continents, megafaunal species began to crash approximately 50,000 years ago, and the wave of extinctions terminated around 10,000 years ago. If we group together Europe, Australia, and North and South America, 101 genera of large mammals went extinct out of 142, which equates to a 70 percent loss (probably close to 90 percent at the species level, although the true number of species is more difficult to assess). It is worthwhile to note that small animals, less than 10 kilograms (about 20 lb.) or so in body weight, do not show any surge in their extinction rate over the same period.

The percentage of megafauna that went extinct is comparable to the overall figures of the largest mass extinctions on record, such as the end-Cretaceous crisis. Besides the magnitude of the extinction wave, its short duration is equally impressive: less than 50,000 years, and perhaps less than 2,000 years on some continents, as we shall see. Either way, the time interval is nearly instantaneous by geological standards. Since the extinctions are so recent, fossils are well preserved and easily accessible, helping paleontologists search for the cause, or causes, of the crisis.

The first obvious explanation is that humankind is to blame for the megafauna's mass extinction. The disappearance of large mammals coincides with the spread of *Homo sapiens* across the globe and the rise of organized hunting, along with the development of elaborate projectile weapons. *Homo sapiens* was officially charged with the mass extinction in 1965, during a meeting of the International Association for Quaternary Research, held in Boulder, Colorado. In a contribution titled "Prehistoric Overkill," American anthropologist Paul Martin made a strong case that humans were to blame.[1]

For the human species to be declared guilty, as in all criminal trials, the prosecution must supply strong evidence, not only because the accusation is extremely serious as such but because it also affects the ongoing debate about the impact of humankind on the biosphere today. In defense of humans, one can argue that the demise of the megafauna also coincides with the chaotic ending of the last glacial stage: after the maximal extent of glaciers around 21,000 years ago, a progressive warming phase began 15,000 years ago, before a brutal relapse of cold weather in the Northern Hemisphere, 12,800 to 11,500 years ago, known as the Younger Dryas. Many paleontologists consider this cold snap to be a good explanation for the megafaunal collapse, which would push the responsibility of humankind into the background.

Hence, since the late 1960s and the initial indictment of humankind in the matter, there has been considerable debate as to who or what is to blame for the demise of the megafauna, and if humankind is indeed responsible, through what precise mechanism. Scenarios range from a progressive decline of animal populations as hunting expanded in step with the rise of human populations (overkill theory) to a more abrupt decline associated with hunting groups surging into virgin territories (blitzkrieg theory). Other correlated causes include humankind's destruction of animal habitats, voluntarily or not, through bushfires and other tampering with nature.

In order to pass judgment, several key questions must be answered. Where were the suspects (i.e., our *Homo sapiens* ancestors) when the extinctions took place? Have we found the murder weapons on site? And if climate was to blame instead, were climatic changes synchronous with extinctions, affecting the right places at the right time? Since we have several crime scenes (different continents) to test our theories, the first place to start is to plot the timing of extinctions against the spread of human civilization across the globe.

The African Case

The first continent to consider—and not the least since it was the cradle of the human species—is Africa. Unfortunately, fossil sites of recent animals are rather rare in Africa, making it difficult to precisely date the extinctions. Paleontologists relied for some time on data compiled in the 1990s that established that from approximately 100,000 to 10,000 years ago, only seven genera of large animals went extinct (including one elephant, one horse, one bison, and one camel genus), joined by three other genera within the past 10,000 years (one buffalo, one bison, and one antelope genus); whereas thirty-eight genera survived, which places the ratio of megafaunal extinction in Africa around 20 percent, a much smaller figure than elsewhere in the world (outside of Africa, the extinction ratio is over 70 percent).

This appears to be somewhat of a paradox: on the continent

where the human species rose to prominence, the megafauna seems to have suffered less than elsewhere. Could this mean that the human species is off the hook from the start, and that some other killing mechanism is to blame, for instance some brutal snap of cooling in the high latitudes, while Africa was spared by its milder climate?

The prosecution can make the case, however, that the situation in Africa is special. One factor to consider is that the human genus coevolved for over 2 million years with the megafauna, establishing an equilibrium between hunters and hunted, large prey getting used to the human threat and developing evasion tactics over time, which saved them from extinction. Another factor is that the human population in Africa might have remained somewhat limited if the coevolution of numerous parasites and bacteria kept mortality high. Our ancestors might have shaken off diseases when they left Africa for colder climes, expanding and becoming a greater threat to the megafauna, because of their larger numbers, than on other continents: an interesting theory but a hard one to prove.

Another possibility yet is that our assessment of extinctions in Africa is incomplete, and the small number of African victims is an illusion. In particular, we might be focusing too much on the period of the last glacial stage, which is the time frame of extinctions on other continents. In Africa, one might need to widen the time interval under scrutiny: many paleontologists, including Paul Martin, have called attention to the fact that the African megafauna could have declined much earlier, and they suggest a first wave of extinctions between 2 million and 1 million years ago that eliminated *Deinotherium*, a giant relative of modern-day elephants; the weird-looking *Ancylotherium*, which resembled a cross between a sloth and a giant goat; and *Homotherium*, the scimitar-toothed predator that disappeared from Africa one and a half million years ago. Paul Martin made the connection between these early extinctions and the rise of the human genus in Africa, in particular *Homo ergaster/erectus*, which probably carried out group hunting at the time. One can thus make the case that in Africa, our ancestors had begun to cut down the megafauna early on, eliminating the most vulnerable species, so that once *Homo sapiens* took over the scene, the megafaunal species left were the most resilient ones, less prone to extinction than on other continents where hunters clashed with the megafauna later, more suddenly, and with a more visible impact.

Another possible misreading of the history of the megafauna in Africa, suggested by paleontologist J. Tyler Faith, is that the number of African species that went extinct recently (over the past 100,000 years) has simply been underestimated. On the basis of recent discoveries, J. Tyler Faith tallies two dozen species of megafauna, rather than one dozen, disappearing during this last time frame; these included the giant warthog *Metridiochoerus*, one giant anteater, one rhinoceros, and several equids and antelopes. Some of them are fossil species newly discovered, others are species that were thought to have gone extinct much earlier but are now found to have lived on as late as the last 100,000 years, when *Homo sapiens* emerged on the scene.

This said, J. Tyler Faith does not conclude that humankind is responsible for their demise, quite the opposite in fact. He points out that most victims were browsers that roamed the grasslands and that changes in the structure and availability of grassland habitats were most probably the cause of their demise. Refining the theory in an article published in 2018, the anthropologist and his coauthors expanded their study to the African megafaunal community of the past 7 million years and made the case that large mammal species began to decline around 4.6 million years ago, before any possible large-scale interaction with our hominin ancestors, and attributed their demise to the climate-driven expansion of tropical grasslands to the detriment of trees and shrubs, which put many browsers in jeopardy, while grazers did somewhat better.[2]

In summary, based on the record in Africa, it is difficult to judge the relative roles of climate and vegetation change, on the one hand, and human hunting and tampering with the ecosystem, on the other, in explaining the demise of megaherbivores. But in all fairness, our human suspect should be discharged, for lack of proof, if Africa were an isolated case.

Megafaunal Extinctions in Europe

If we turn to the situation in Europe, the record is also complex, in particular because at higher latitudes, climate oscillations during the Ice Age were much more pronounced than on the African continent. Glaciers advanced and retreated in step with the forays of human hunters, making it difficult to untangle the relative roles of climate and hunting.

The European ecosystem was remarkable in those days. From the Atlantic coast to the easternmost reaches of Siberia, a vast corridor of tall grasses, mosses, and lichen unfurled between the glaciers to the north and the forests to the south, a corridor best described as a "steppe and tundra" environment. Each time the climate cooled, the corridor expanded as forests retreated southward; during warmer interglacials, the band became narrower as the forest line advanced northward. This steppe and tundra corridor was a remarkable environment in which northern and southern animals mixed, especially during the summertime, when grasslands were comparable to Africa's Serengeti National Park today: an environment that hosted an odd mix of woolly rhinoceroses, hippopotami, wild pigs, leopards, mammoths, elephants, bison, and hyenas. The proportions of different species fluctuated as their populations expanded out of their geographical niches when conditions were favorable and retreated back to their reserves when the climate hardened: cave bears preferred western Europe; mammoths and woolly rhinoceroses were more sensitive to the cold than one would imagine, leaving Great Britain and Germany to head south as soon as temperatures plummeted; and aurochs and large deer stuck to the forest lines.

As elsewhere on Earth, the term "megafauna" is certainly appropriate for this European menagerie: cave bears (*Ursus spelaeus*) weighed up to three-quarters of a metric ton (1,600 lb.), double the weight of brown bears today. Giant elks (*Megaloceros giganteus*) had antlers that spanned over 4 meters (13 ft.) and weighed up to half a metric ton (1,100 lb.). Ice Age elephants (*Palaeoloxodon antiquus*), most likely related to today's African elephant, weighed up to 11 metric tons (24,000 lb.): their fossils are found from Great Britain to Spain, across the Netherlands, Germany, and France, and across Asia, as far east as Japan. As for the celebrated mammoth, its fossil bones are found all across Europe, including four nearly complete skeletons in France—one of them, discovered in 2012, east of Paris, is aged close to 100,000 years and linked to a couple of flint stones found near its skull. Although the tools certainly belonged to Neanderthals, it is not clear if they were used to hunt down and kill the animal or simply to skin and butcher it after it died of some other cause.

The human genus lived among this European menagerie as early as 600,000 years ago (*Homo heidelbergensis*). Bones of an elephant dating back to 420,000 years ago were discovered in Great Britain, near the river Thames, again connected with nearby flint tools. Direct proofs of hunting, attributed to later Neanderthals, appear somewhat later: an elephant skeleton aged 120,000 years, discovered near the town of Lehringen in Germany, bears a yew-wood spear lodged between its ribs. At a time when the human population was certainly sparse, connecting one of its weapons directly with a megafaunal casualty is significant.[3]

Once *Homo sapiens* appeared on the European scene, around 40,000 years ago, it quickly displaced its Neanderthal cousin and took over the European continent. Was the newcomer better armed? Was it better clothed against the cold or better organized? Or did it simply come in larger numbers? Whatever the reason, not only did *Homo sapiens* drive the Neanderthals out of existence—our invading species also arguably exerted an unprecedented pressure on the environment. That said, in order to be held responsible for the collapse of the European megafauna, it must be shown that the appearance of *Homo sapiens* in Europe coincides closely in time with the extinction of most large animals. Is this really the case?

Tallying extinctions is a difficult task. On the one hand, dating techniques are difficult to implement and often imprecise. On the other, one must always keep in mind that the youngest fossil of a species only marks a provisional date for its extinction: the possibility always exists that an even younger specimen will one day be found. A case in point is the European elephant, long believed to have disappeared over 50,000 years ago, before *Homo sapiens* arrived on the scene. In 2007, however, fossil bones were discovered in the Netherlands, estimated to be under 40,000 years old, and a wall painting in Vermelhosa, Portugal, apparently depicting the head of an elephant, suggests that the species survived in Europe until around 30,000 years ago.

Three other large mammals disappear from Europe in the time frame stretching from 40,000 to 15,000 years ago, which corresponds to the expansion of *Homo sapiens* across Europe: the two-horned interglacial rhinoceros (*Stephanorhinus*), the cave bear (*Ursus spelaeus*), and the great spiral-horned antelope (*Spirocerus*). The final, crucial stage of megafaunal extinctions in Europe is 15,000 to 7,500 years ago, during which time the woolly rhinoceros (*Coelodonta antiquitatis*), the cave lion (*Panthera spelea*), the woolly mammoth (*Mammuthus primigenius*), and the giant elk (*Megaloceros giganteus*) went extinct.

One might argue that brutal cold snaps, accompanied by the expansion of land ice, which occurred both 25,000 and 12,000 years ago, shook up and exterminated these animals, with humankind playing no significant role in the process. However, considering that the interglacial rhinoceros arrived in Europe 600,000 years ago and the woolly mammoth at least 170,000 years ago, these species and others endured earlier cold spells that were equally brutal. The difference is that they went extinct during the specific glacial stage when *Homo sapiens* arrived in Europe.

Climate swings are hard to incriminate, without implicating a human accomplice, for another reason. On the scale of an entire continent, like Eurasia, species troubled by a climate swing would be expected to migrate until they found a suitable haven, weathered out the bad period, and expanded back out, once the crisis was over. The fact that they disappeared altogether suggests that they were harassed by some other factor within their very havens.

Two well-documented cases of extinction shed light on the issue. One is the fate of the giant elk, *Megaloceros giganteus*. Based on fossils found in Ireland, it was long believed that the last members of the species went extinct in Eurasia around 12,800 years ago, at the onset of the last cold snap of the Ice Age, known as the

Younger Dryas. However, new fossil discoveries and better dating indicate that the species survived elsewhere until 7,700 years ago, past the cold crisis, ending up in a restricted area of Ukraine and western Siberia. A team of paleobiologists made the case that, on the one hand, the elk's diminishing presence across Europe corresponds to the narrowing of its favored feeding habitat—mixed steppe-woodland—as the cold spell intensified and that subsequent warming was of no help, promoting new forests unsuitable to the animal's diet. On the other hand, note the experts, even if there is no archaeological proof of giant elk hunting by humans, the disappearance of the animal from Ireland, around 12,800 years ago, corresponds to the first signs of human presence in the island; in the Ural region, the presence of humans is attested 8,000 years ago, narrowly predating the final extinction of the giant elk, 7,700 years ago. As the authors of the study conclude, the climate swings "reduced giant deer populations to a highly vulnerable state. In this situation, even relatively low-level hunting by small human populations could have contributed to its extinction."[4]

The second well-documented case is the extinction of the woolly mammoth, *Mammuthus primigenius,* long believed to have gone extinct 12,000 to 11,000 years ago. Imagine the surprise of Russian naturalists when they discovered mammoth remains that were only 4,000 years old, on Wrangel Island, off the shore of Eastern Siberia. No climate event corresponds to this final extinction date; the date matches the late arrival of human hunters on the remote island.

To conclude, if we try to make sense of all the evidence concerning megafaunal extinctions in Europe, the gradual disappearance of species appears to match the expansion of *Homo sapiens* across the continent, but it is also closely tied to climate swings that were much more severe than in Africa. One can then imagine that climate crises repeatedly shrank megafaunal populations into dispersed, small pockets of survival. Under normal circumstances, one would expect the animal populations to expand back out of their refuges once the climate improved, as appears to have been the case several times during early stages of the Ice Age. It can be argued, then, that during the last stages of the Ice Age, human hunters tracked the vulnerable populations down to their very refuges, committing local massacres that turned into absolute extinctions if these populations were the last representatives of the species. According to this scenario, *Homo sapiens* could be charged for the extinction of megafaunal species in Europe, but with mitigating circumstances provided by the underlying stress of climate change. We should keep this in mind when it comes to the present wave of extinctions in which humankind is involved: adverse factors might affect species that otherwise would rebound were it not for the added stress of human activities that break the camel's back, so to speak—be it overhunting, habitat reduction, or climate change.

The Americas Take the Stand

Whereas the megafaunal extinction leaves us with no clear culprit in Africa and shared responsibility between humankind and adverse climate in Europe, the Americas provide key evidence that incriminates the human species.

North America provides well-dated sites of megafaunal fossils and a precise calendar for the arrival of *Homo sapiens* on the continent: although there a few older signs of human presence, the major immigration wave of future Americans across the Bering Straits, during the last stage of the Ice Age, is believed to have started around 15,000 years ago. Coming over from Siberia across the shallow, iced-over straits, the migrants rapidly dispersed across the Americas. Evidence of campsites stretches from Washington State to Virginia and Delaware, and down through Arizona, New Mexico, and Texas. Uncovered artifacts include projectile points carved out of flint, chert, or jasper, fluted at the base where they were attached to wooden spears or darts. Named the Clovis culture, in reference to a prominent archaeological site near the city of Clovis, New Mexico, this early American civilization flourished over a time span of four centuries, from approximately 12,900 to 12,500 years before present.

Around this time, which coincides with the final cold snap of the Ice Age (the Younger Dryas), the Clovis civilization faded out,

The Hypothesis of Celestial Intervention

In October 2007, one year after launching the idea in a popular science book, nuclear chemist Richard Firestone and geologist Allen West were joined by twenty-four other scientists in raising the hypothesis of a cometary impact, 12,900 years ago, that would have caused the Younger Dryas cooling event—and allegedly the associated megafaunal extinctions. The evidence they presented is the occurrence on many sites in North America, dating back to this period, of lake-bed "black mats," containing soot, bits of charcoal, and an abnormally high concentration of iridium, platinum, and metal, as well as silica-rich microspherules, which the authors attribute to cosmic impactors, the high-temperature blasts, and the wildfires that resulted.[5]

Iridium and other metals common in meteorites have also allegedly been found in ice cores in Greenland dating back to the same period, and the number of sites displaying impact evidence is climbing, extending from dozens of locations in North America to other locations in South America, Europe, and the Middle East, namely at Abu Hureyra in Syria—an archaeological site marking human transition from a nomad hunting settlement to an agriculture-based civilization.

Compiling and checking the evidence, chemist Wendy Wolbach, a combustion-carbon and wildfire specialist who previously worked on the end-Cretaceous mass extinction (see chapter 2), summarized with her coauthors in a 2018 paper that "the Younger Dryas interval is one of the most unusual climate episodes in the

retreating to a restricted, more central area of North America, and evolved so much in tool style and other cultural aspects that archaeologists refer to it as a different civilization, which they call the Folsom culture. Some archaeologists believe that the transition from Clovis to Folsom was brutal, rather than gradual, and invoke catastrophic causes, such as the impact of an asteroid or comet somewhere in North America (see "The Hypothesis of Celestial Intervention").

Whatever the cause of its ultimate downfall, the Clovis civilization clearly marks the first large-scale occupation of North America by *Homo sapiens*, with archaeological sites extending to Mexico and South America. In fact, one of the oldest Clovis sites was discovered in Mexico's Sonoran Desert, at a fossil site named El Fin del Mundo: "the end of the world." Dated 13,400 years before present, five centuries before the golden age of the Clovis entire Quaternary record," and that "a cosmic impact is the only known event capable of simultaneously producing the collective evidence."[6]

An independent review by chemist Martin Sweatman in 2021 agreed that the case was convincing, stating that "the scale of the event, including extensive wildfires, and its very close timing with the onset of dramatic Younger Dryas cooling suggest [that clues of an impact event] are plausible and should be researched further."[7]

The proposed scenario is that of a collision between the Earth and a cloud of cometary debris circling the Sun, with a number of fragments exploding during their entry into the atmosphere. Airbursts, similar and probably much larger than the Tunguska event of 1908 over Siberia, stretched over a third of the circumference of the Earth, setting off wildfires and arguably triggering, via soot darkening of the atmosphere or other mechanisms, the abrupt Younger Dryas cold snap and perhaps related events, such as the local decimation of wildlife and a shake-up of human communities and cultures.

On the other hand, many experts have attempted to check and reproduce the data and have not been able to confirm most of it, casting doubts on the hypothesis—a controversy further muddled by pseudoscientists joining the debate. It will take time to sift through the real and the erroneous data, check and crosscheck, but in the end, truth will prevail.

culture, this early site gives a clear indication of the hunting threat directed against the megafauna: four lance points, carved out of translucent quartz, were found amid the fossil bones of two gomphotheres (*Cuvieronius*), elephant-like mammals that belong to the list of great animals that went extinct around that time.[8]

Five centuries later, New Mexico's Clovis type site (12,900 years before present) reveals hundreds of lance points, associated this time with no less than fourteen piles of mammoth remains, numerous marks of butchering on the bones, and even one lance point found directly lodged in the shoulder blade of a bison. Several other Clovis sites are equally replete with hunting or butchering bone beds of mammoths and mastodonts (the elephant proboscidean family)—a surprisingly large number for a time interval lasting no more than five centuries, so much so that paleontologist Gary Hanes stresses that these proboscidean kill sites are *"far more in number than on any other continent and far more in such a brief time interval than in all of prehistory,"* Africa included. Hunting of large animals by the Clovis people appears therefore to have been particularly intense.[9]

One recent discovery is particularly startling. In the soft lake-bed sediment at White Sands National Park in New Mexico, a team of scientists spotted hundreds of tracks left by giant sloths and human beings alike, loosely dated sometime between 15,000 and 10,000 years before present, an interval that includes the time frame of the Clovis culture. Human footprints lie within the sloth tracks, indicating that they belonged to hunters stalking their prey; after some distance, sloth trackways show abrupt changes of direction, as if the animals were trying to escape, and deep imprints even suggest the sloths reared on their hindlimbs in defense, while concentrated human footprints indicate a terminal episode of harassment. The tracks at White Sands thus constitute exceptional evidence of humans hunting down large preys in North America.[10]

In summary, the fact that the wave of immigration of human hunters in North America coincided with the decline of the American megafauna is quite convincing. The major wave of immigration, as we recall, began around 15,000 years ago. In Wisconsin, a first site of mammoth bones with butchering marks dates back to this period. From this first proof of human hunting to the end of the Clovis era—a span of 2,500 years—fifteen genera of large animals went extinct, and another twenty disappeared broadly around the same time (but their extinction dates are still too imprecise to be counted as reliable data points). If we limit ourselves to the fifteen genera extinctions that are well dated, an interesting point is that the more these extinction dates are refined, the more they converge toward the five centuries at the end of the 2,500-year bracket when the Clovis culture reached its apex.

There is another, indirect way of tracing the decline of the American megafauna, which is particularly ingenious. Ice Age ecologist Jacquelyn Gill called an unlikely witness to the stand: *Sporormiella*, a microscopic fungus that develops on the dung of herbivores, including mammoths. By counting the number of fungus spores across layers of lake-bed sediment (Appleman Lake in Indiana), spanning the period from 17,000 to 8,000 years before present, Gill and her team brought to light a spectacular drop in the concentration of *Sporormiella*, starting approximately 15,000 years ago, coinciding with the arrival of *Homo sapiens* in North America. According to the fungal proxy, the population of large mammals reached its lowest value 13,700 years ago, never to spring back. All megafaunal species had not yet gone extinct at the time, since fossil bones are still found up to the 12,000-year mark, but the 15,000-to-13,700-year bracket seems to frame the collapse of their populations.

The lake-bed sediment also records the evolution of vegetation over the same period, through the concentration and makeup of fossil pollen. A sharp change occurs *after* the collapse of the megafauna, in the form of a surge of black ash and hophornbeam trees, rapidly joined by pine and oak. Clearly, since the change of vegetation *follows* the extinctions, rather than precedes them, it is not the cause of the extinctions. One can even argue that vegetation change is a *consequence* of megafaunal collapse, the disappearance of the large browsers releasing the pressure they exerted on certain types of trees.[11]

Since there is no sign of a climate change in this string of events; the most obvious agent of megafaunal extinctions in North America boils down to *Homo sapiens*. As a corollary to this hypothesis, the decline of the Clovis culture might be because the hunters drove to extinction their very source of food, compounded by the need to adapt to a changing landscape.

Climate change does jump in at the very tail of this succession of events, and it might have contributed to the change from the Clovis to the Folsom culture. The Younger Dryas cold snap, which took place from 12,900 to 11,500 years ago, probably resulted from a change of ocean currents in the North Atlantic. It did not trigger the extinction of the megafauna, since it kicks in more than 1,000 years after the population began to collapse, but it is quite possible that the surviving animals were driven by the cold snap into ever-shrinking habitats, where they were easily finished off by human hunters.

The fact that the climate was not the principal cause of megafaunal extinction in the Americas is further confirmed by the record in South America. Glacial stages significantly affected the northern continent but much less the southern one; yet in South America, megafaunal extinctions are as drastic as they are in North America, if not more. No less than fifty species go extinct, and only ten that survive—hence an extinction rate of 80 percent. Famous victims include the giant sloth (*Megatherium*), close to 6 meters (20 ft.) long and weighing 3 metric tons, along with its smaller cousins; members of the elephant family, such as *Stegomastodon* and *Cuvieronius*; browsers of the camel family, including the strange-looking *Macrauchenia*; *Hippidion* and other horse-related species; and large carnivorous species, best represented by the saber-toothed *Smilodon*.

As was the case in North America, no change in vegetation is noted at the beginning of their demise, hence no proof of a climate crisis. By contrast, megafaunal extinctions clearly coincide in time—at least those extinctions most precisely dated—with the arrival of *Homo sapiens* in South America 15,000 years ago (immediately following its colonization of North America), according to camp sites in Chile and Patagonia. Hence, at this time, humankind remains the only suspect in the megafaunal extinctions of South America, although research in the matter is still in its early stages.

Australia Called to the Stand

Humankind appears guilty of megafaunal extinctions in North and South America, though much less so in Europe and Africa. The testimony of Australia appears crucial to settle the issue.

For starters, Australia had fewer genera of large mammals (sixteen) than the Americas, but it had the largest number of casualties: fourteen genera disappeared out of sixteen: an extinction rate of nearly 90 percent. All the great Australian browsers bit the dust: the giant wombat *Diprotodon*—the largest marsupial to have ever existed at up to 3 metric tons in weight; the giant flat-faced kangaroo; and marsupial carnivores, like *Thylacoleo carnifex*, comparable in weight to female lions today, with a catlike head, powerful jaws, and a tail like a kangaroo's. In fact, the only large mammals still alive in Australia today are the wombat genus and one kangaroo genus (*Macropus*).

The extinctions are spread out over the past 80,000 years, but half are dated with enough precision to fall inside a narrow 10,000-year-bracket, 50,000 to 40,000 years before present. It so happens that *Homo sapiens* landed in Australia around 55,000 years ago, and it was well established on the continent 45,000 years ago, based on the first camp sites. This is precisely the time period when many megafaunal genera went extinct.

As was done in North America, one can use the *Sporormiella* fungus as a proxy to trace the decline of Australian mammals over time. Sampling of sediment on the lake bottom of Lynch's Crater, Queensland, shows a sudden drop of *Sporormiella* starting 41,000 years ago, arguably the sign of megafaunal collapse, coinciding with the spread of human hunters. There is no observed climate crisis in Australia that matches the extinctions. On the contrary, climate changes on record that precede the extinctions, including two events occurring 75,000 and 55,000 years ago, had no apparent effect on the megafaunal population. Besides the *Sporormiella* proxy, the research team that studied the crater lake sediment, led by paleoecologist Susan Rule, pointed out a noticeable rise of charcoal debris that begins roughly a century *after* the collapse of the megafauna. The surge in charcoal is followed a couple of centuries later by a long-lasting change in the makeup of the vegetation, transitioning from lush forest and savanna to small-leaved, chaparral-type bushes and trees (sclerophyll vegetation), typical of more arid conditions.[12]

Altogether, the evidence in Australia suggests the following sequence of events. The rise in numbers of *Homo sapiens* hunters in Australia, 45,000 to 40,000 years ago, caused the collapse of the megafauna, setting off a chain reaction. The disappearance of the megafauna upset the balance of the entire ecosystem: without large herbivores browsing the trees, dry wood accumulated and wildfires became more frequent, clearing the forests and setting off a vicious circle in which the land became drier at the scale of the entire continent. Unless contrary evidence is discovered,

Homo sapiens thus appears to be the root cause of the megafaunal extinction in Australia, and of the ensuing climate change.

The Verdict

With the evidence presented for five continents—Africa, Europe, North and South America, and Australia—it is now time to come to a verdict regarding the main culprit or culprits of the mega-faunal mass extinction, but not before a final witness is called to the stand, in defense of *Homo sapiens*.

In a related development, geologists and chemists unearthed evidence that the Earth encountered a cloud of cometary debris 12,900 years ago, allegedly sparking a string of airbursts over several continents—most notably North America—that blasted and charred the ground, set off wildfires, and lifted large amounts of soot into the atmosphere, allegedly triggering the Younger Dryas cold snap (see "The Hypothesis of Celestial Intervention").

This hypothesis, first presented in 2007, is now losing ground, since much of the data presented as proof is being challenged and does not stand up to closer scrutiny.

However fascinating, the Younger Dryas impact event and cold snap occurred 12,900 years ago—long after the extinction of most megafaunal species. Therefore, these events are acquitted of the murder of the megafauna on the grounds of this lack of synchronicity, except perhaps for the extinction of a few of the last species—and even for those, there is no conclusive evidence.

This latest development aside, research over the last two decades appears to exonerate climate and vegetation change as a principal cause of megafaunal extinction, except in Europe where it could have played a supporting role. In the Americas and Australia, *Homo sapiens* remains the main suspect, accused of having overhunted large mammals and reduced their population to levels below which reproduction was no longer possible.

Reproduction is indeed an important factor in the survival or demise of animal species. In the case of the end-Cretaceous mass extinction, as we recall, large animals like the dinosaurs were the hardest hit; no animal larger than 25 kilograms (55 lb.) survived the crisis. The reason most commonly cited is that large animals are already self-limited in numbers, meaning that a catastrophic drop of their population ultimately reaches a critical level below which survivors do not interact enough to reproduce efficiently.

During the Ice Age megafaunal collapse, the same mechanism probably played out, with human hunters acting the part of the asteroid impact. Another reproduction issue is that large animals reproduce at a slow rate. Whereas a female rabbit, for example, averages four or five litters per year, each litter providing half a dozen bunnies, the pregnancy of a female elephant lasts twenty to twenty-two months and results in only one baby. If large animals with long gestation periods are hunted down and their females slaughtered, they cannot reproduce fast enough to make up for casualties, leading to a catastrophic collapse of their populations.

In a statistical analysis of victims and survivors among Ice Age mammals, Christopher Johnson, professor of wildlife conservation, concluded that the probability of extinction of Ice Age species exceeded 50 percent—in other words, it became more probable than not—if their reproduction rate was less than one newborn per adult female per year. Exceptions that survived, noted Johnson, were species that lived in Arctic or alpine environments or had an arboreal or nocturnal way of life. This is somewhat enlightening, in that it appears to further exonerate climate change as a cause of megafaunal extinction: if the climate—namely a cold snap—were to blame, in what way would an arboreal or nocturnal way of life have offered protection? On the other hand, with respect to the *Homo sapiens* hunting scenario, one can speculate that arboreal or nocturnal behavior hid these species from hunters who operated mostly during the daytime, in the open savanna.[13]

Two other observations, more general in character, appear to exonerate climate in the extinction of the megafauna. The first one is that the megafauna developed and flourished during the full extent of the Ice Age, spanning over 2 million years, and survived over a dozen cooling cycles, including the Günz, Mindel, and Riss glacial stages, as they are named in Europe (known as the Nebraskan, Kansan, and Illinoian stages in the American Midwest). The wave of extinctions only occurred during the last glacial stage—the Würm, or Wisconsinan, glaciation—which was no more severe than the previous ones but corresponds to the worldwide expansion of *Homo sapiens.*

The second observation is that small populations of megafaunal animals survived exceptionally late, strikingly in places that were isolated from human civilization and where the climate was no different from elsewhere. The best-known case is the extended survival of mammoths on the remote Taymyr Peninsula in the Far North of Russia, and on the uninhabited islands of Wrangel (off the coast of northeastern Siberia) and Pribilof (off the coast of Alaska). These late mammoths shrank in size in response to their limited habitat—a natural process known as insular dwarfism—and lived until 4,000 years ago (as opposed to 12,000 years elsewhere), when it appears that hunters finally managed to reach their remote locations.

The giant elk *Megaloceros* is also believed to have found refuge in isolated environments, like the Isle of Man (between Ireland and Scotland) and the Ural Mountains, where it survived several thousands of years beyond the main population hunted down by *Homo sapiens*; it finally succumbed 5,000 years ago. The same observation can be made for the giant sloth, which first disappeared from the mainland of the Americas over 10,000 years ago but survived in the Caribbean islands until 5,000 years ago, its final extinction coinciding with the first evidence of *Homo sapiens* reaching its last refuge.[14]

Faced with several converging lines of evidence, it seems difficult to deny that *Homo sapiens* played a major, if not a dominant role, in the crash of the megafauna, between 50,000 and 10,000 years ago. As pointed out, other factors probably played a role, such as climate instability or even cometary impacts, but they appear circumstantial rather than the primary culprit.

The geological record also shows how the environment reacted to this first episode of global change caused by our species. The sudden decline of large grazers, which fed mainly on grasslike vegetation, led to unconstrained grass growth, which in turn sparked a sharp increase in wildfires, signaled by charcoal spikes in the sediment.[15]

The last glacial episode ended 11,650 years ago, a warming trend that set the most recent stage in the history of the Earth, known as the Holocene. As its Greek name indicates (*holos* for "whole" and *kainos* for "new"), the Holocene represents a whole new episode in the history of the Earth: the ongoing interglacial stage that saw the rise of human civilization.

5 Extinctions across Recorded History

While a few questions still remain as to the exact role of humankind in the disappearance of the megafauna, there is no doubt about the direct responsibility of human civilization in the next wave of extinctions, which followed during the Holocene. First there was a pause. The most vulnerable, larger animals had already been exterminated, so that their decline no longer stood out. As for smaller species, even if their populations were falling in response to overhunting and habitat loss, their existence was not yet threatened. The negative pressure exerted by *Homo sapiens* on biodiversity returned to the forefront when our species began to colonize oceanic islands and their closed ecosystems, cornering the few remaining species of megafauna that had survived the first onslaught.

It is worth recalling the testimony of Wrangel Island, twice the size of New York's Long Island, 140 kilometers (90 mi.) off the coast of Northern Russia in the Arctic Ocean. This is where numerous remains of small-sized woolly mammoths were discovered, the most recent ones aged 4,000 years—the date of their final extinction. Hence, they were able to survive 6,000 extra years on the island, with respect to continental mammoths. They reached Wrangel Island by drifting or swimming (their elephant cousins are known to be good swimmers), and their shrinking in size—insular dwarfism—was caused by the limited food available and the absence of predators, which rendered useless the protection offered by large body size. The respite ended 4,000 years ago. The extinction of Wrangel's woolly mammoth coincides with the first traces of human presence on the island. Since there was no change of climate or vegetation at the time, human hunters are left with the sole responsibility of the animal's demise.

Similarly, several species of giant sloths temporarily survived in the Caribbean islands, long after vanishing from the mainland of the Americas. The *Megalocnus, Parocnus, Acratocnus,* and *Neocnus* genera, totaling twelve different species, survived on the islands of Cuba and Hispaniola (which is present-day Haiti and the Dominican Republic) until around 5,000 years ago, with again no sign of climate change in the Caribbean during their demise, whereas the first traces of human occupation are synchronous with their extinction, followed by a drop in the amount of tree pollen and a rise of wood charcoal in the sediment, pointing to the burning and destruction of the environment. Directly through overhunting, or indirectly through slash-and-burn behavior, *Homo sapiens* stands out as the main culprit of a wave of extinctions that took only 1,000 years to kill off all the giant sloths in the Caribbean archipelago, as well as four monkey species and three-quarters of all insectivorous mammals in the islands.[1]

Madagascar and New Zealand

Two other examples of local extermination are particularly well documented and cover a much larger number of species: Madagascar and New Zealand.

The island of Madagascar in the Indian Ocean, covering an area comparable to that of France or Texas, is separated from the coast of southeastern Africa by straits no more than 400 kilometers (250 mi.) in width (the Mozambique "canal") but sufficient to isolate it from human intervention until recently: the first evidence of human hunters on the island dates back only 2,000 years. From this time onward, six species of giant birds, also known as "elephant birds," vanish from the scene. Larger than ostriches, they were hunted down for their eggs. The last of these flightless birds went extinct in the late seventeenth century. It stars in a short story by H. G. Wells: "Aepyornis Island" (1894).

The onslaught was not limited to giant birds. In the same 2,000-year interval, seventeen species of lemurs went extinct in Madagascar, accompanied by two species of hippopotami, two species of giant tortoises, one large anteater, and one carnivorous mongoose-like mammal. In summary, all Madagascar animals larger than 10 kilograms (roughly 20 lb.) in weight have been exterminated since the arrival of *Homo sapiens* in the island.

For this unfortunate fauna, as was the case in the Caribbean, no climate upheaval or major change in plant life can be invoked to exonerate humankind as the principal cause of extinction. On the contrary, a detailed study of recent sediment layers in the island indicates that a change in plant spore, as well as spikes in charcoal attributed to wildfires, *follow* rather than precede the wave of extinctions, as was the case in North America and Australia during the demise of the megafauna. Vegetation change did not cause the downfall of animal species; it was the extinction of species that led to a change in vegetation—a deterioration that was certainly aggravated by further human interactions with the environment.[2]

Following the spread of human civilization across the oceans, the wave of extinctions that ran from island to island next hit New Zealand. Isolated from Australia by a stretch of ocean 1,500 kilometers (930 mi.) wide, New Zealand was colonized by Polynesian navigators (the Māori people) in the early 1300s. New Zealand hold the dubious distinction of being the last region on Earth to have its megafauna massacred, at a time, again, when no climate change can be blamed for the destruction.

New Zealand's star victim was the moa bird, the skeletons of which were discovered by European navigators and their teams of natural history experts when they first explored the islands at the end of the 1700s. The moa was impressive indeed: building on 60 million years of evolution, this unique lineage of giant flightless birds numbered nine different species when the Māori reached the islands, the largest couple of species standing 3 meters (10 ft.)

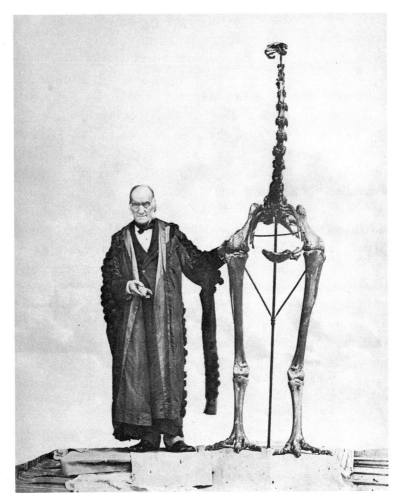

Paleontologist Richard Owen poses next to a reconstructed skeleton of the extinct moa bird. From Richard Owen, *Memoirs on the Extinct Wingless Birds of New Zealand*, vol. 2 (London: John van Voorst, 1879), plate XCVII.

tall and weighing over 200 kilograms (440 lb.): a third taller and heavier than ostriches. Totally wingless—an evolutionary trait that resulted because they had no predator on the island (except for one local eagle species)—the moa birds were not used to any kind of threat, and therefore they were an easy target for Māori hunters. Hunted down for their meat and their eggs, the moa birds were reduced to piles of bones throughout the islands.

As with other recent extinctions, there was at first some reluc-
tance to indict humankind. Volcanic eruptions in New Zealand
were suggested as a cause of the moa's demise—the islands are
famous for their spectacular volcanoes—and even a cosmic impact was invoked, inspired by the blast that hit the Siberian forest
at Tunguska in 1908 and set off wildfires around ground zero.[3] A
parallel can also be drawn with the dinosaur extinction debate, in
that several researchers claimed that the moa birds were declining
prior to the arrival of the Māori people in the early 1300s and that
hunters only provided a final blow to an animal genus that was already on the verge of extinction.

More work in the field has proven otherwise. A research team,
led by evolutionary biologist Morten Allentoft, extracted genetic
material from the bones of three hundred specimens from four different species of extinct moa birds—a sampling covering a 12,000-
year time interval—and concluded that there was no decline in
diversity or population over the last thousands of years of their
existence: moas were doing fine until their sudden collapse. Paleontologists estimate that moas numbered close to 60,000 individuals when Māori hunters reached the islands in the early 1300s.
Carbon-14 dating indicates that the brunt of extinction took place
within 200 years of the onset of human presence. Interestingly,
it appears that small populations of moa birds found refuge in
remote alpine areas of the South Island (around Fiordland) and
survived an extra couple of centuries before their final hunt-down
and extermination in the late seventeenth century. This supports
the concept of refuge zones where last pockets of populations can
temporarily survive, as was the case for the megafauna of Wrangel
Island, the Isle of Mann, and the Caribbean.[4]

In New Zealand, moas were not the only victims. Other large
birds disappeared on the islands as well, in just as short a time
span, including the Haast's eagle—the largest eagle species to
have ever existed, weighing an estimated 15 kilograms (roughly
30 lb.)—which probably fed on young moas (and, therefore, was
likely a collateral victim of the moa extinction); two local, large-
sized goose species, weighing over 40 kilograms (90 lb.); two
carnivorous birds (*Aptornis*, also known as adzebills), featuring

stout legs, atrophied wings, and a large beak; and six duck species. These large birds made up the totality of the megafauna in New Zealand, since there were no large reptile or mammal species on the islands. Besides the fact that New Zealand was the last scene on Earth of the wave of extinctions that hit the megafauna, the islands also suffered a concurrent extinction of smaller animals, credited to inconspicuous but ferocious accomplices of human migrants.

When the Māori people landed in New Zealand, they brought with them little furry stowaway animals: ship rats. The rodents began to decimate all the smaller birds that Māori hunters over-looked, by eating their eggs—New Zealand birds nested on the ground, since they had no natural predators. The situation got even worse when European sailors began colonizing the islands in the late 1700s, bringing along not only unwanted rats but also cats, hunting dogs, and even stoats and ferrets. Reproducing at a fast rate, these newcomers massacred not only birds but also small amphibians and reptiles, and one of three local species of bats, the only endemic mammals of New Zealand.

One might never know how many unique species of lizards and geckos were eradicated, but a few made history, in particular, *kawekaweau*, the giant gecko. In Māori lore, the giant gecko represented the soul of deceased relatives, and it was also a bad omen for whoever laid eyes on it. This might have been why, in 1870, the last individual was killed by a Māori chief, and the slain animal was never put on display. Paleontologists actually believed the giant gecko to be a myth until the 1980s, when Alain Delcourt, curator of the Natural History Museum of Marseille in southern France, accidently came across a 60-centimeter (2 ft.) stuffed gecko in the museum's basement and circulated a photograph of the animal. It wasn't long until reptile specialists formally identified the animal as being the legendary *kawekaweau*: a specimen probably collected by French explorers in the early 1800s. The giant gecko became the logo of the Marseille Museum, and the species was named after its curator Alain Delcourt: *Hoplodactylus delcourti*.[5]

While we might never know how many reptile species were exterminated by *Homo sapiens* in New Zealand, the good news is that several dozen species of lizards and geckos did survive on smaller islands off the main coast, islands that were bypassed and spared by explorers and their armies of dogs and rats. But for how long?

If we turn our attention back to New Zealand birds, the official record shows fifty-one species gone extinct since the islands were settled, including thirty-eight bird species exterminated by Māori hunters and nineteen exterminated by European explorers and settlers. This figure could be higher, however, since paleontologists keep discovering other recently extinct species, notably by combing through owl feces that contain bones of small birds that were once part of their diet and have since disappeared.

The problem with assessing extinctions is that we fail to tally species when there is no record of them before they went extinct. This is especially true for small species for which fossil bones have passed unnoticed, such as small birds, reptiles, and amphibians, not to mention invertebrates and plants. Conversely, some species have left a lasting impression, because they were remarkable, attractive, or even revered.

Such was the case of the huia, an elegant bird endemic to New Zealand admired for its dark blue plumage highlighted by greenish iridescence; it also had an orange wattle (a fleshy outgrowth) under its beak and a white stripe at the end of its tail. Huia feathers were prized by the Māori people, which contributed to the species' decline, but the bird's doom was ultimately the deforestation and habitat destruction caused by European settlers, as well as an ancillary but devastating hobby: bird collecting.

The role of collectors, be they butterfly hunters or amateur collectors of stuffed birds, is often overlooked when it comes to population drops and extinctions. The pressure on a species becomes all the more vicious because the rarer a species gets, the higher the price for a specimen, leading to more intense hunting. There was a time when preserving animal populations was not a major concern, and museums of natural history abused the ecosystem just as much as individual collectors. In the case of New Zealand's beautiful huia bird, museums were paying large sums to obtain stuffed specimens for display, contributing to the extinction wave. One taxidermist based in Austria held the gruesome record of killing 212 couples of huia birds in 10 years' time, all for Vienna's Natural History Museum. As an aside, it is interesting to point out that collectors' passion for birds goes beyond the grave: during an auction in Auckland in June 2010, a huia feather sold for NZ$8,000, the highest price ever paid for a bird feather.

The last huias ever to be observed alive were a group of three birds spotted in York Bay (near Wellington) in December 1922. Two other questionable reports signal one or two birds near Lake Waikareiti in 1961 and 1963. More than 50 years have elapsed since these last sightings, so that the species is now officially deemed extinct, or to be honest, exterminated.

If we translate all these losses into statistics, the current estimate is that since the arrival of *Homo sapiens*, New Zealand has lost 40 percent of its bird species: an extinction figure that can most likely be extrapolated to all New Zealand vertebrates as a whole. Though it is spectacular, the case of New Zealand is far from unique. It stands as a sobering example of the wave of extinctions sweeping across the globe since our species has taken over the world, birds representing the living symbol, or rather the dying symbol, of our impact on nature.

How to Do Away with Birds

The great voyages of discovery, which began in the 1400s and peaked in the 1700s and 1800s with James Cook and Charles Darwin, had two major and opposite effects. The first one is that they led to the discovery of new species, helping Darwin in particular to establish his theory of natural selection and evolution, based on differences that the naturalist noted between birds along the coast of South America and, most famously, in the Galápagos Islands.

The second effect is that, concurrently, explorers and their followers destroyed many of the very species they were discovering, by invading their fragile habitats and exploiting them without restraint, for food or for the sake of collectors. At the same time that human beings were discovering the laws of evolution through observation and theory, they were also discovering the concept of extinction, as they drove the process, albeit unwittingly, into practice.

Islands are fantastic reservoirs of biodiversity, full of original species that evolved independently from continental ecosystems. But most insular species are particularly vulnerable, because they have been exposed to few predators and find themselves powerless when hunters land with their companion animals. All islands have experienced this calamity, but one example is most telling in this regard: Mauritius Island and the fate of one of its native species, the dodo bird.

East of Madagascar in the Indian Ocean, Mauritius Island is less than 2,000 square kilometers (800 sq. mi.) in area—comparable in size to the Hawaiian island of Maui—and was not settled by *Homo sapiens*, apart from short visits by Arab and Portuguese sailors, until the landing of Dutch explorers in the late 1500s. One of its endemic inhabitants was the dodo (*Raphus cucullatus*), a fat bird belonging to the pigeon family, weighing up to 25 kg (55 lb.)—comparable to a large turkey—and incapable of flying; it was a convenient food stock for hungry navigators. Cats and dogs contributed to the dodo's demise, as did deforestation and habitat destruction. The massacre was carried out at lightning speed. The first dodo on record was reported by a Dutch sailor in 1598; the last dodo was shot by a hunter in 1688. The species was exterminated in less than a century, leading many naturalists in Europe to question whether the bird had ever existed at all, or if it was merely a legend invented by sailors, despite sketches and paintings that seemed to prove otherwise.

That such a remarkable bird disappeared in such a short amount of time seems to defy common sense, raising the uncomfortable notion of extinction. It was reassuring to believe that the animal was inherently "stupid" and poorly adapted to its environment, the same argument once used to explain the demise of dinosaurs. In fact, a famous expression was coined—"go the way of the dodo"—to mean "become obsolete."

Contrary to these simplistic beliefs, dodos, like dinosaurs, were well adapted to their environment. It was an external agent of unusual violence—an asteroid for dinos, human beings for dodos—that eliminated from the face of the Earth a species that certainly did not "deserve" to perish any more than any other.

The fact that a rare species or two, with a small number of individuals, might perish in fragile insular environments when faced with human beings desperately in need of food could be written off as insignificant collateral damage of the great voyages of discovery, without implying that extinctions might strike well-stocked species on a continental scale. The destructive potential of *Homo sapiens*, however, is brutally illustrated by the dramatic fate of the passenger pigeon.

Native to North America, the passenger pigeon (*Ectopistes migratorius*)[6] was famous for assembling into huge flocks that stunned observers as the bird continuously migrated, breeding around the Great Lakes and flying as far south as Florida and the Gulf of Mexico to feed and winter. Numbering anywhere from three billion to five billion in its heyday, the passenger pigeon might well have been the most numerous bird species in recorded history. One particular day in 1866, in southern Ontario, a flock representing perhaps the entire population flew overhead: it was estimated to have been 1.5 kilometers (1 mi.) wide and 500 kilometers (300 mi.) long, taking fourteen hours to cross the sky, which translates into an estimated three and a half billion birds.

Less than 50 years later, on September 1, 1914, the world's last passenger pigeon died in captivity: a new speed record for a human-caused extinction, all the more shocking in that the species spanned an entire continent and comprised several billion members. How could *Homo sapiens* have exterminated such a prolific species in such a short amount of time, without the assistance of pollution or climate change?

Simply through hunting. Passenger pigeons represented a threat to farmers, devastating their crops; and they were also a cheap source of meat to feed the enslaved population and the poorer cast of city people. Their migratory behavior conveniently distributed the fowl across a large area, from Canada to Florida, in particular along the Great Lakes, where they congregated in forests of white oak, often numbering several hundred per tree. A hunter could fire a buckshot into a tree and kill fifty pigeons at

The North American passenger pigeon (*Ectopistes migratorius*) went extinct after industrial hunting in the nineteenth century. From *Illustrated Sporting and Dramatic News*, 1875.

a time. Some hunters simply chopped down trees, or set fire to them, to drive pigeons to the ground. At the same time, large-scale deforestation occurred in North American to clear land for agriculture.

The pigeon-meat market exploded when cities were supplied by railroad. Pigeons became hunted on an industrial scale, and great nets were spread across the land that could trap up to thirty-five hundred pigeons at a time. One report states that in the spring of 1878, in Petoskey, Michigan, fifty thousand pigeons were killed *every day* over a span of five months. Do the math, and you will end up, for one village, with a toll of seven million birds in less than half a year. While we might shudder at the thought of over two hundred huia birds killed in New Zealand by one Austrian taxidermist, how can that compare to the deaths of *three million* passenger pigeons credited to a single Michigan hunter in 1878?

From millions to billions, there is only one step left to reach extinction, but it's a big one. Why did some small fraction of the pigeon population not overcome the crisis, since its drop in numbers undoubtedly made the large-scale hunting techniques less profitable and pigeon-meat too expensive for the market? Wouldn't the scarcity of the bird become its main ally?

It appears that a fatal threshold came into play, which led to the final collapse of the pigeon population. It probably had to do with the bird's way of life. Migration involved large numbers of individuals, and arguably so did reproduction: large-scale mating behavior only took place, perhaps through some type of hormonal response, when birds reached a certain concentration in numbers. Whatever the precise mechanism, once the population dropped below a certain level, reproductive efficiency, or survival behavior at large, collapsed: a vicious circle that led to the extinction of the passenger pigeon at lightning speed.

Hence the sad end of the story. In the 1880s, when hunters and protectors of the environment realized that the pigeon population was crashing, it was already too late. By 1900, there were no passenger pigeons left in the wild. Despite last ditch efforts to rescue the bird, including protecting several groups of individuals in zoos and museums, the last official individual of the species, a female named Martha, died on September 1, 1914, at the Cincinnati Zoo in Ohio. A statue was erected in her memory, a stark reminder of human-caused extinctions. Biologist Paul Ehrlich sums up the drama by stating that the passenger pigeon's extinction "illustrates a very important principle of conservation biology: it is not always necessary to kill the last pair of a species to force it to extinction."[7]

The Status of Birds Today

As these historical examples so tragically illustrate, birds are often the first casualties of the extinction onslaught. They offer us an opportunity to size up the extent of the destruction caused by humanity and to assess whether extinction rates are comparable with those of the five great mass extinctions that have previously struck the biosphere.

Birds are close cousins of the dinosaurs, and they are the only branch of their class that made it through the end-Cretaceous mass extinction. Like the mythical phoenix reborn from its own ashes, birds bounced back from the crisis to repopulate the Earth. Could birds crash again but this time disappear from the face of the Earth?

Birds are particularly well studied. It is estimated that over 95 percent of all bird species alive today have been identified, which provides a basis for meaningful statistics. The tally reaches a grand total of 11,188 extant bird species cataloged as of January 2023, according to the International Union for Conservation of Nature (IUCN), with half a dozen new species added to the list each year. The second figure of importance is the number of species that officially went extinct during human history—that is, those that were seen alive and described by naturalists but have now totally disappeared. The number of these extinct bird species stands at 160, as of January 2023.[8]

The extinction percentage is simple to derive by dividing the second number by the first: approximately 1.5 percent, which is a significant figure but does not yet seem to be an alarming one compared to the greater than 50 percent extinction figures reached during a mass extinction. But some perspective is in order. First, if we look at the geographical distribution of these historical bird extinctions, there are blatant zones of crisis. New Zealand is first, with a toll of 60 lost species, followed by the Hawaiian Islands with about 30 losses. These include coots and other members of the rail family (which are mostly ground-living birds), a number of passerine birds that fed on flower nectar, and one crow species. The fact that New Zealand and Hawaii are the worst-hit sites comes as no surprise: birds there had narrow habitats and were particularly attacked by rats and other predators that came off the boats, including the sailors and settlers themselves. Hence, if we focus our count on island birds only, the extinction figure reaches 8 percent—far above the world-average of 1.5 percent. By contrast, birds in large continental areas took longer to be threatened, and only recently did continental species begin to disappear, with three losses recorded in the 1800s and ten losses during the 1900s, including the famous passenger pigeon.

Besides regional discrepancies, the second cause of concern is that the rate of bird extinctions has picked up over the last several decades. Approximately 20 species of birds went extinct between 1975 and 2000, and a similar number will disappear by 2025. Notable losses include Maui's yellow and green 'ō'ū forest bird (*Psittirostra psittacea*), hit both by habitat loss and avian malaria, as well as the Maui nukupu'u (*Hemignathus affinis*), a yellow-breasted honeycreeper, and the po'o-uli (*Melamprosops phaeosoma*), a black-faced honeycreeper—its last representative dying in captivity in 2004.

The bottom line is that one bird species goes extinct every one or two years on average. In order to put this figure in perspective, it is important to compare it with the "natural" rate of extinctions for birds—that is, the average background value, exempt of human interference, that we can establish by looking at fossils of bird species and the time bracket each one spans. A broad average is that a bird species "lasts" approximately 1 million years before going extinct. If we scale this figure to the roughly 10,000 species of extant birds, it translates into a probability that one bird species should naturally go extinct every century or so. Since human civilization kills off a bird species every *one to two years*, our rate of massacre appears to be fifty to one hundred times higher than the natural background rate.[9]

How will this translate over time if our killing rate doesn't change? Adding 50 casualties per century to the present total of 160 extinct bird species gives us 10,000 years to go before we eliminate 50 percent of the avian biodiversity and turn the current crisis into a great mass extinction. Optimists will feel reassured by this projection and could argue that we have ample time to fix the problem, with no need to change our way of life.

The situation, however, might be far worse. We don't know if we really have tallied all the bird species that we have driven out of existence. The figure of 160 extinctions is largely based on observations by naturalists over the past five centuries, but how many other species were destroyed in prior centuries, when Polynesian sailors and hunters hopped from island to island? There is no description of what they found or what no longer exists, and even when naturalists did start recording species, some birds might have been dying off outside of their line of sight. A dying bird is much more difficult to spot than a flock of live ones that sing, fly, and end up getting painted.

Ecologists Richard Duncan, Alison Boyer, and Tim Blackburn estimated that nearly 1,000 nonpasserine land birds went extinct and unrecorded across the Pacific. This is based on the bones of unrecorded specimens discovered on several well-studied islands, with the figures then extrapolated to the entire ocean basin. These "ghost birds" suggest that instead of the extinction of 160 bird species since the spread of human civilization, the total could exceed 1,160 species.[10] Based on this number, the rate of extinction over the last millennium would then be about 12 percent, seven to eight times higher than the official 1.5 percent.

If we adopt this figure of 12 percent, augmented by one bird extinction every couple of years, we reach the 50 percent figure of a great mass extinction a little sooner than in our previous estimate: in 8,800 years rather than 10,000 years. Still far into the future, optimists will say. The situation, however, really depends on one final factor, and it is not the least: Will birds continue to go extinct at the present rate, which is already fifty to one hundred times the natural rate, or will the pace of extinctions quicken? This is the challenge faced by our civilization today. We will explore in the next chapter what we might expect based on current trends and projections.

The Extinction of Mammals

The status of mammals is of particular concern to us, since we belong to this very successful—at least until now—class of animals. A number of mammal species play an important role in our everyday lives, be they livestock or pets, and even wild species of mammals cause us to feel tenderness and empathy, since we share common traits, such as social behavior and care for the young.

One might expect that all mammals in existence have long ago been cataloged, but we keep discovering new species at a surprising rate: while only 3,000 mammal species were known in 1900, the figure now reaches 6,615 species as of January 2023, which represents a doubling in numbers over the past century.[11]

Most new mammals that make the list are not well-hidden unicorns or yetis; most are small rodents, bats, or the like, many of which were long considered to be subspecies, before being promoted to full species status. Others that join the list are surprisingly large. A new antelope variety, named Walter's duiker (*Philantomba walteri*), was described in 2010. It long escaped notice because it is extremely rare, hunted in Togo and Nigeria for its meat and horns; in fact, this particular duiker has not been observed in the wild by scientists: only a few killed specimens, spotted in bushmeat markets, led to its identification.

In 2011, it was an astonishing little monkey that came out of the hat: living in mountain forests on the border of China and northern Burma, this snub-nosed colobine (*Rhinopithecus strykeri*) is very rare but unexpectedly easy to spot, because its unprotected nostrils cause it to sneeze when it rains. In 2012, a new monkey species was identified, this time in the Democratic Republic of Congo. Much more discreet than the colobine, this long-nosed lesula (*Cercopithecus lomamiensis*) has shy, humanlike eyes that make it particularly endearing. Equally cute is the tree-living olinguito (*Bassaricyon neblina*), a nocturnal mammal of the raccoon family, with big brown eyes and a face like a teddy bear, discovered in South America's montane forests in 2013. And the search goes on: one more monkey made it on the list in 2020—the Popa langur (*Trachypithecus popa*) of Burma—followed in 2021 by a cream-colored lemur with orange hues (Schneider's marmoset, *Mico schneideri*), spotted in the Amazon rainforest in Mato Grosso, Brazil, and a maned sloth with a coconut-shaped head (*Bradypus crinitus*), also from Brazil, registered as a new species in 2022.[12]

Unfortunately, most of these recently discovered species have few individuals, and many are threatened with extinction, which brings us to the less rejoicing tally of mammals that did go extinct on our watch—a tally that begins officially in the year 1500 CE, when we have a sufficiently good record of described species, and of those that disappear thereafter, with the understanding that this figure is a minimum, since we might discover skeletons of mammals that disappeared before they were officially described, as we saw to be the case for island birds.

Two comments are in order here. First, a number of large mammals went extinct during the expansion of human civilization in prehistorical times: the megafaunal crisis at the end of the last glacial stage, roughly 50,000 to 11,000 years ago, which we covered in the previous chapter and detached from the ongoing crisis. One question for future paleontologists to work out is whether extinctions attributed to *Homo sapiens* should be defined as consisting of two separate waves, or bunched together as one. A second point is that few mammals occupied the Pacific islands (except for a few bats), as opposed to birds, so that when Polynesian and European navigators landed in these ecosystems, the tally of mammal extinctions did not surge, as it did for avian species.

Be that as it may, 80 mammal species have gone extinct since the 1500s—half of them over the past 100 years, either because the census methodology improved in recent times or because extinctions have truly accelerated. Against a total of approximately 6,600 assessed mammal species (as of 2023), this amounts to a loss of 1.2 percent, only slightly less than the extinction figure for birds.

With respect to the geographical distribution of losses, Australia was most affected: its original cast of mammals was especially vulnerable to hunting and invasive species, and it was further hit by deforestation and intensive agriculture. Over two dozen endemic species disappeared in recorded history—more than 10 percent of Australian mammals. Marsupials were the hardest hit: half a dozen kangaroo and wallaby species went extinct during the twentieth century. The most famous casualty has to be the thylacine (*Thylacinus cynocephalus*), better known as the Tasmanian tiger or Tasmanian wolf.

This original marsupial resembled placental carnivores on other continents—with a fox-like head and a striped back, recalling tiger markings—but differed from them by its marsupial

The Tasmanian wolf (*Thylacinus cynocephalus*) was hunted down by farmers, who mistakenly believed it attacked their sheep. The last known individual died in Hobart Zoo in 1936. From National Archives of Australia: A1200, L35618.

pouch and kangaroo-like tail. The thylacine lived in the grasslands and eucalyptus forests of mainland Australia, New Guinea, and Tasmania; a nocturnal hunter, it preyed primarily on birds and rodents, but it was mistakenly singled out by farmers and herders as a threat to livestock. The decline of the thylacine began over 40,000 years ago on the Australian continent, during the megafaunal extinction allegedly linked to the arrival of the human species. In its last stronghold of Tasmania, the thylacine's final demise was accelerated by bounties offered by landholders to exterminate the animal. Over two thousand were officially massacred between 1888 and 1909, and probably many more off the record. By the time the thylacine had almost disappeared in the 1920s, the Tasmanian Advisory Committee for Native Fauna attempted to save the species, although the motivation was most likely to provide zoos with thylacines, rather than protect the animal for its own sake. The effort came too late: the last known individual died on September 7, 1936, in Tasmania's Hobart Zoo during a cold night when it was left outside as a result of neglect. Just as the passenger pigeon became the symbol of extinctions in the United States, the death of the last thylacine is now commemorated every year on September 7: Australia's Threatened Species Day.

How to Become Officially Extinct

Keeping a tally of mammals that have gone extinct (or any animal or plant, for that matter) is not an easy matter. Not too long ago, it was customary to declare a species extinct if none of its members had been spotted over a period of 50 years. In the case of the thylacine, reports of sightings lingered on for over half a century since the last reported death in 1936, but none have been confirmed. Unless an individual shows up—and what a surprise that would be—the thylacine has thus officially been extinct since 1986 (the 50-year cautionary interval, since the clock started ticking in 1936).

Record-keeping rules, however, are evolving fast. As data collecting is becoming more efficient, and the issue of extinction more pressing, the cautionary time lag between the last spotting of a species and declaring it extinct has been brought down from 50 to 20 years. Zoologists and nature watchers have become more reactive: when a species appears to be on the verge of extinction, expeditions are mounted to spot the last individuals and save them if possible. Such was the case of China's freshwater baiji dolphin (*Lipotes vexillifer*), which deserves a special mention, since it stands as the last large mammal to go extinct, since the Caribbean monk seal in the 1950s.

Sporting a pale blue back, a white belly, and an upturned beak, nicknamed "the goddess of the Yangtze River," the baiji suffered from its very aura, when Chairman Mao Zedong launched his Great Leap Forward economic plan (1958–62), criticizing "backward" folklore and the veneration of animals, and encouraging instead their exploitation and hunting. Already hard-hit by fishing nets and boat traffic, the baiji population rapidly collapsed. Only a few hundred individuals remained in the 1970s, and although hunting the dolphin was finally banned in 1983, the construction of large dams across the Yangtze, in particular the Three Gorges Dam, brought a final blow to the species, further restraining its motion and polluting the river. In 1997, a dedicated international expedition to sight the baiji came up with only thirteen individuals spotted. In 2004, only one baiji was reported, prompting Chinese and international observers to view the situation as critical and mount a new expedition. In 2006, a six-week cruise, covering 3,500 kilometers (2,200 mi.) up and down the river, failed to spot a single baiji.[13]

Besides monitoring the alarming situation, the Chinese government did work on a conservation effort in the 1990s to save the baiji and other freshwater dolphins: an oxbow lake alongside the Yangtze River was turned into a reserve and proved partially successful, since it protected another endangered species—the Yangtze finless dolphin (*Neophocaena asiaeorientalis*)—but the one baiji placed in the oxbow in 1995 died the following year.

For the baiji, conservation efforts came too late. In December 2006, after the dedicated cruise that failed to spot a single individual, the species was declared "functionally extinct": even if two or three baijis escaped notice and are still alive today, that number would be insufficient to save the species. Efforts should now be turned toward saving the finless dolphin; although less than a thousand individuals are left, the population decline seems to have reversed in the late 2010s, as a result of China's conservation efforts.

Reptiles in Trouble

Evolving from lobe-finned fish, amphibians were the first vertebrate animals to set foot on dry land, 370 million years ago, further splitting into other classes—reptiles (sauropsids) and protomammals (synapsids)—a few tens of millions of years later.

In the public mind, amphibians and reptiles are often regarded as having fallen behind in the struggle of evolution—some are even described as "living fossils"—as if only birds and mammals were worthy of respect. Despite their seemingly archaic features, however, amphibians and reptiles are success stories in their own right. If their aspect has changed little over time, it is not that they represent the backwaters of evolution; on the contrary, it is that they are so well adapted to their environment that they do not need to change much.

Reptiles, in particular, are more numerous than one would tend to think. As their census has progressed, the number of species broke the 10,000 mark in 2014 and reached 12,000 in 2023, neck and neck with birds for second place in the vertebrate world (behind fish), and way ahead of mammals. Specifically, there are over 7,000 species of lizards, over 4,000 snakes, 363 turtles, and 27 crocodilians.[14]

With respect to extinctions, reptiles appear to have fared better than birds and mammals, with about 40 species reported to have gone extinct in recorded history, which corresponds to approximately 0.3 percent. Recall New Zealand's *kawekaweau*: as was the case for this giant gecko, most reptile extinctions affected island species, which were massacred by cats, dogs, and other predators that landed with explorers and settlers. The list of casualties includes the Martinique curlytail lizard (*Leiocephalus herminieri*), last collected in the 1830s; Guadeloupe's giant ground lizard (*Pholidoscelis cineraceus*), last recorded in 1914; and the Round Island burrowing boa (*Bolyeria multocarinata*), endemic to islets surrounding Mauritius, in the Indian Ocean, last seen in 1975. One of the last announced victims—a subspecies, rather than a species—was a giant Galápagos tortoise (*Chelonoidis niger abingdonii*). Although its last official representative, a male named "Lonesome George," died on Pinta Island on June 24, 2012, a female relative was spotted and captured on the neighboring Isabela Island in 2020. Attempts will be made to breed this female with males of another subspecies in order to save its lineage.

Many more reptiles are in the process of being added to the extinction list, not so much because they noticeably went extinct over the past few years but because experts are catching up on species that have not been spotted for several decades, and the time has come to declare them extinct. The blue-dotted gecko of

The last individual of the Pinta Island (Galápagos) giant tortoise (*Chelonoidis niger abingdonii*) died in 2012. After enduring three centuries of hunting and habitat destruction, the subspecies has been declared extinct. Photo by putneymark, licensed under CC BY-SA 2.0.

Mauritius, a skink lizard native to Réunion Island, and Australia's Christmas Island forest skink are but a few.

Figures should also be reassessed upward for the same reason cited for island birds: many reptiles went extinct before being described and cataloged, and their very existence escaped scrutiny until excavations revealed their bones. One study in particular targeted the Guadeloupe Islands of the French Caribbean and unearthed thousands of bones, spanning a time interval of 40,000 years: it revealed that eight species of reptiles—mostly lizards and one snake—existed up to and throughout colonization of the islands by pre-Columbian settlers and went extinct in step with the arrival of European colonizers: they were likely exterminated by cats, mongooses, and other predators that traveled with the colonizers, as well as by deforestation linked to new agricultural practices.[15]

The bottom line is that over 50 percent of reptile species in Guadeloupe went extinct before records even began, and if this figure is extrapolated to other islands, it is not 0.3 percent of reptiles that have disappeared worldwide since the beginning of civilization but a much larger figure.

The Plight of Amphibians

As was the case for reptiles, the diversity of amphibians was long ignored. Since the start of the twenty-first century, their number is skyrocketing, with an average of 160 new species discovered and cataloged each year, amounting to 10 percent more species added to the list every decade. Their total number broke the 8,500 mark in 2022, of which 7,500 are frogs and toads, close to 800 are newts and salamanders, and the last 200 or so are limbless, snake-like creatures, called caecilians.[16]

Some new species are truly astonishing. The record of the smallest vertebrate in the world was clinched by an amphibian, discovered in 2013: a frog endemic to New Guinea, named *Paedophryne amauensis*, that is only 7 millimeters (0.25 in.) long. This surge of newly discovered amphibians eclipses by far those that have gone extinct under our watch over the past 500 years: only 38 amphibian species were listed as having gone extinct as of 2022.[17]

As research progresses, however, the figures are quickly changing, and there is reason for concern. The first observation is that amphibian extinctions are on the rise. Out of the three dozen species that disappeared in the last 500 years, over a third did so very recently, in the past 50 years. The trend is worsening, since the discovery of new species in remote and fragile locations often shows them to be in a critical state, to the point where scientist helplessly watch their ongoing extinction.

Such was the case of the extraordinary platypus frog of Eastern Australia, also known as the gastric-brooding frog (*Rheobatrachus silus*), so-named because the mother swallowed her fertilized eggs and embryos, incubated the tadpoles in her stomach, and regurgitated them as young frogs. Discovered in 1972, that amazing little animal was last seen in 1981; apparently, a pathogenic fungus accidently introduced into its native habitat caused its demise.

The second, related, reason amphibians are suffering more extinctions than official figures suggest is that their census is difficult, costly, and underfunded. Yet, as research and expeditions proceed, there is an alarming "death row" of species that are currently being investigated and are about to be declared extinct. In 2023, the number of extinct amphibians was probably much closer to 200, rather than the official 38. That would bring the percentage of extinctions in the amphibian world from 0.4 percent to over 2 percent, which is a much more alarming figure.

The third point of concern, already mentioned in the case of birds and reptiles, is that many amphibians were driven to extinction by the rise of civilization without leaving a trace of their demise, unless their bones are discovered. For every one that is unearthed by chance and pulled out of oblivion—such as the mummy of a Sri Lankan frog discovered in the drawer of London's Natural History Museum—how many will forever go unnoticed? Overall, the current extinction figure for amphibians is by no means trivial, and as we will see in the next chapter, the group's future does not look promising.

The Fate of Fish

After mammals, reptiles, and amphibians, what do fish have to tell? Considering that life began and evolved in the oceans, one would expect marine animals to be the most numerous and diverse species, when compared to land-living beings. Surprisingly, the opposite is true. Land insects, plants, and mushrooms outnumber them by far, and the largest marine group—mollusks—comes only in fourth position among the main groups of lifeforms, on par with arachnids. It counts, nonetheless, around 100,000 cataloged species (including a minority of terrestrial mollusks).[18]

Crustaceans, including shrimp, crab, and barnacle species, clinch sixth position, with over 50,000 cataloged species, and only in eighth position do we find the fish group, with close to 35,000 species identified as of 2023, both freshwater and marine. Fish do remain, however, the most diverse group of vertebrates, far ahead of birds, reptiles, amphibians, and mammals.

The census of marine and freshwater lifeforms is far from over, since approximately 700 species of crustaceans, 500 species of mollusks, and 300 species of fish are added to the list each year, along with about 100 new coral species (the cnidarian group, totaling about 10,000 species), 50 new sponges (added to a total of 6,000), and 30 starfish and urchins (echinoderms, totaling 7,000 species). Summing up all the groups, our water world boasts approximately 250,000 species: a number that rises by about 2,000 new members each year (an annual growth rate of close to 1 percent) in our ongoing census.

Recent members that joined the club include a sea anemone (*Edwardsiella andrillae*) that lives anchored upside down to sea ice along the shores of Antarctica, with two dozen tentacles extended downward to catch plankton; a candelabra-shaped sponge (*Chondrocladia lyra*), living 3,400 meters (11,000 ft.) below sea level, off the coast of California; and the surprising pancake batfish (*Halieutichtys intermedius*) of the Gulf of Mexico, which drags its flattened body across the seafloor on its fins, like a bat walking on land.

The tally of marine species is far from over, especially since it relies on expensive technology and complex oceanic surveys, involving both surface vessels and submersibles diving deep below the surface. According to specialists, the number of marine species that await discovery is probably close to a million: a large figure but still far from threatening the numerical advantage of terrestrial species.

If we now turn to extinctions, the record is surprisingly modest. One would have thought that overfishing, climate change, and pollution would have caused notable casualties over the years, but such is not the case, at least for the time being. In recorded history (since the year 1500), only 92 species of fish have officially gone extinct, although the list might soon gain 130 more casualties, listed as "possibly extinct species." Overall, however, these figures represent a loss of diversity of only 0.2 to 0.6 percent.[19]

When we pore over the list of deceased species, there comes a second cause of astonishment. Only one seawater fish is officially extinct: the smooth handfish *Sympterichthys unnipenis*, which used to live in the coastal waters off Tasmania.[20] All others are freshwater species, and there is a reason for that. In the oceans, fish can swim away from unfavorable environments—like polluted, overheated, or overfished waters—and migrate toward zones of refuge, even if large numbers of individuals die in the process. Freshwater environments, on the other hand, are often closed, limited systems. Any stress brought to this type of environment, be it overfishing, pollution, climate change, or the introduction of an invasive species, can be fatal. Among freshwater fish that have gone extinct, many lived in small ephemeral ponds, critically dependent on rainfall. Repeated droughts ended up killing entire populations, as happened in Arizona to *Cyprinodon inmemoriam*, a member of the "pupfish" group, which owes its nickname to the behavior of males during mating season: charging and fighting each other on the sand bottom like puppy dogs at play. As for the species name, *"inmemoriam"* says it all.

Larger freshwater habitats can also harbor extinctions, a fate that befell the thicktail chub (*Gila crassicauda*), a large minnow that once roamed the marshes and slow-flowing rivers of California's Central Valley. The chub was overfished, and its population was affected by water diversion schemes to irrigate crops in the Central Valley, until it finally disappeared altogether: the last individuals were caught in the 1950s, and the only place to see thicktail chubs today is in the glass jars of local museums.

Besides overfishing and habitat destruction, another major threat to freshwater fish is the introduction of invasive species into their habitat. A famous case is the Nile perch (*Lates niloticus*), a large predator that was introduced in many African lakes to support the fishing industry. Individuals can grow up to 200 kilograms (440 lb.) in weight, and they help support over two hundred thousand fisherpeople — up to two million people, if one includes ancillary jobs. The downside is that the Nile perch feeds on native fish wherever it is introduced — Lake Victoria being the most famous example — and is held responsible for the ongoing extinction of hundreds of local species. When the death toll becomes official, it will be too late.

On other sites, extinctions brought about by invasive species have already gone down on record. Such was the case of New Hampshire's silver trout (*Salvelinus agassizii*), overwhelmed by a perch that was introduced by recreational fisherpeople in the early 1900s; the Moroccan trout of Lake Aguelmame Sidi Ali

(*Salmo pallaryi*), stamped out in the 1930s by the introduction of the common carp; and Lake Titicaca's flat-headed pupfish (*Orestias cuvieri*), exterminated by nonnative trout in the 1950s.

While fish extinctions are rather anecdotal for the time being, and no world-famous species has disappeared from market stalls yet, the situation might evolve significantly in decades to come.

Shellfish and Crustaceans

Crustaceans count many more species than fish—approximately 50,000 versus 35,000—but display even less recorded extinctions in historical times: only a dozen or so in official listings. Again, most extinct species are freshwater animals, with the marine world still off the hook, so to speak.

One freshwater shrimp species disappeared in Florida and two small crayfish from Mexican ponds, devastated by excessive irrigation and downward collapse of the water table. California scores two casualties: a freshwater shrimp that lived in rivulets around Los Angeles, its habitat destroyed by urban sprawling, and a local crayfish (the sooty crayfish, *Pacifastacus nigrescens*) around San Francisco, eradicated by the signal crayfish (*Pacifastacus leniusculus*), a more common, robust species introduced by recreational fisherpeople. Incidentally, California's signal crayfish is one of three invasive species introduced in European waters that are decimating the populations of local crayfish. The latter are affected by a disease carried by the introduced crustaceans, known as the crayfish plague: a fungus that is harmless to its North American carrier but deadly for European species that come into contact with it. The bottom line, however, is that despite the decline of European crayfish—locally a popular food stock—fish market stalls are still replete with shrimp, crab, and lobster, and crustaceans do not appear to be a prime source of concern, at least not yet.

More at risk are mollusks: a rich animal group that comprises around 100,000 described species, such as marine clams and oysters, and terrestrial snails and slugs. As of 2021, the IUCN listed 300 species that had gone extinct since the 1500s (roughly 270 gastropods and 30 bivalves), which is already significant, but the figure is probably grossly underestimated, according to a study published in 2017. Checking with experts, biologist Robert Cowie and his group came up with over 600 extinct mollusk species and another 400 that are possibly extinct, which could push the total over the 1,000 mark: triple the official number. Casualties are mostly land snails (78 percent), followed by freshwater snails (17 percent), freshwater bivalves (4 percent), and only half a dozen marine gastropods.

The report focuses on two case studies, the first one on a "test" sample of 200 cataloged land snails around the world. Out of the approximately 120 species for which experts had enough data to render a judgment, over 10 percent were evaluated as having gone extinct (30 species). This extinction rate is an alarming figure that mirrors similar generalizations made for birds, reptiles, and amphibians. The second study was conducted in Hawaii on one group of endemic land snails, known to be in decline: the Amastridae family, once numbering 325 known species. Of this representative sample, 131 species were found to be extinct, and 179 were impossible to assess but possibly extinct as well. Only 13 snail species were found to be alive. Although this family of Hawaiian land snails might not be representative of mollusks or invertebrates as a whole, it reinforces the previous study and draws attention to the seriousness of the situation.[21]

As for the reasons of such a heavy toll among mollusks, it is the same story as for other land animals: overexploitation, habitat destruction, and invasive species. In French Polynesia's Gambier Islands, it was deforestation; in Romania, it was tourism that caused thermal waters to be overexploited and puddles to dry up; in New Zealand, it was the opening of a coal mine that destroyed a snail's habitat. Hawaii, for its part, offers the sobering example of invasive species: both an introduced predatory snail (*Euglandina rosea*) and invasive rats combined to exterminate several species of tree snails. Even gastronomy plays a part in the attacks on mollusks: the famous *escargot de Bourgogne*—the snails of fancy French restaurants—is experiencing a population crash, as demand for the delicacy increases. There is also the habit of collecting shells, which endangers the most decorative species.

The bottom line is that the death toll of mollusks is underestimated. This is partly because keeping tabs on mollusks is a staggering challenge: although there are roughly as many specialists of mollusks as experts of other animal groups, such as mammals and birds, mollusk specialists have to survey ten times more species than their colleagues. There is a lot of work to be done, and it has to be done now.

Insect Decline

While tracking mollusk species is a daunting task, tallying insects is an even greater challenge. They are the most diverse group of animals on Earth, topping a *million* recorded species, with a growth rate of about 10,000 new species identified each year. Experts estimate that the number of insects still awaiting discovery and classification might reach anywhere between 5 million and 10 million species. Among those cataloged, the beetle order Coleoptera comes first, with 400,000 described species, followed by Hymenoptera (sawflies, ants, bees, and wasps), Diptera (flies and mosquitoes), and Lepidoptera (butterflies and moths).

Recently discovered species that made the headlines in the insect world include a small parasitic wasp, *Kollasmosoma sentum*, flying in squadrons close to the ground and attacking columns of ants from behind to inject their eggs into their helpless targets, and a yellow millipede from Tanzania (*Crurifarcimen vagans*), otherwise known as the wandering leg sausage on account of its evocative shape, that reaches 15 centimeters (6 in.) in length.

To this vast universe of true insects, add the Arachnida class—spiders, scorpions, mites, and ticks—which tops 100,000 species, with 1,500 new species added to the catalog each year. New members of the club include the striking, iridescent blue tarantula (*Pterinopelma sazimai*), which lives in the highlands of Brazil, and a dark blue giant scorpion (*Heterometrus yaleensis*) discovered in Sri Lanka's Yala National Park.

Understandably, with thousands of new species to describe and catalog each year, entomologists have little time to devote to the flip side of the coin: insect extinctions. The official Red List of Threatened Species, compiled by the IUCN, lists only 60 extinct insects, and another 90 or so possibly extinct, based on 12,000 species assessed so far by the organization. This would amount to only a 0.5 to 1.3 percent extinction ratio so far among insects, but such a low figure is partially due to the lack of time and resources available to study the situation in depth.

If we take a look at the list of registered victims, the sampling bias is obvious: seven defunct insects are reported in the well-studied Hawaiian Islands and three in California, confirming the impression that the number of extinct species is directly proportional to the amount of attention granted them, and that deep in the Amazon basin and Indonesia's forests, there are less environmental associations and experts at work than in San Francisco or Honolulu.

While there are few insect extinctions on record, three constitute interesting examples—one because it shows that a large population does not necessarily protect a species from extinction, and the other two because they show the complex interaction and interdependence of species within an ecosystem.

With respect to population collapse, the extinction of the Rocky Mountain locust (*Melanoplus spretus*) is just as spectacular as the fate that befell the passenger pigeon. Throughout the nineteenth century, swarms of locusts were known to sweep through the Great Plains, east of the Rockies, every five or six years on average, from Canada to Texas, devastating crops and "eating everything but the mortgage," as one farmer said wryly at the time. One somber day of July 1874, in particular, a swarm of locusts descended from the Rocky Mountains onto the prairies, eclipsing the Sun in a mobile, milky cloud the size of an entire state. Their numbers were estimated at several *trillion* insects, and they sounded like a hailstorm as they dropped to the ground and piled up to form a thick blanket. Their total mass was estimated at 25 million metric tons, the equivalent load of one hundred super oil tankers: the largest concentration of insects ever recorded, according to the *Guinness Book of Records*.[22]

Their crops completely destroyed, over one hundred thousand farmers were driven to starvation and had to be rescued by the US Army. Less than 30 years later, in 1902, the last Rocky Mountain locust was spotted and captured in southern Manitoba, Canada. Not a single individual has been observed since, and the species was finally declared extinct by the IUCN in 2014.

How could an insect species that had set a world abundance record have gone extinct in such a brief interval of time? Here, again, as with the passenger pigeon, the answer has to do with reproduction. Reaching staggering numbers in the summertime, the locusts dropped to a minimal population in the fall, retreating to their breeding grounds in a small number of valleys at the foot of the Rocky Mountains, along the border of Montana and Wyoming. Female locusts laid their eggs in the ground, but that very ground was upset when ranchers brought their cattle to graze in the valleys, and they replaced the natural habitat with alfalfa crops. Without knowing it, ranchers destroyed the locust's sanctuary. Not only did the stamping of cattle crush the locusts' nests and their larvae but it seems that the alfalfa the locusts fed upon affected their growth cycle in a negative way. It is also possible that beaver hunting dismantled dams along the local rivers, flooding alluvial plains and drowning the larvae. Altogether, this shows how a prolific species can be exterminated if it is struck at the weakest point of its life cycle.[23]

Two other examples are less spectacular but illustrate how species are interdependent within an ecosystem. Such is the case of a blue butterfly, *Glaucopsyche xerces*, that lived in the sand dunes near San Francisco and sustained a symbiotic relationship with a local ant species. The butterfly's caterpillar secreted nourishing honeydew for the ants, and ants in exchange protected the caterpillar from parasites. The delicate balance was upset when the wave of urbanization spreading around San Francisco disturbed the soil, eradicating lotus plants that caterpillars fed upon and also bringing with it an army of ordinary ants, which replaced the local ant species and apparently were not interested in pursuing a partnership with the caterpillar. Losing both its food stock and its ally, the butterfly vanished: it was last seen in 1943.[24]

One last example is the American chestnut moth (*Ectoedemia*

Mushrooms and Other Fungi

One much overlooked branch of lifeforms is the vast group of fungi—mushrooms, yeasts, and molds—which are separate from the animal, plant, protozoan, and protist worlds, and usually hide in soil and dead organic matter. Only about 150,000 species of fungi have been described so far, but their true number probably stands between 2 million and 4 million. Thanks to efforts by mycologists around the world, around 2,000–3,000 new species are cataloged each year, but the number of described species still hovers around 5 percent of the estimated total, not the least because fungi are usually so difficult to find, as mushroom hunters know well.[25]

Fungi were cataloged, but considering the difficulty in assessing their status, they were long ignored in extinction counts and lists of threatened species. The first couple of fungus species—lichen fungi—did not appear on the IUCN's Red List of Threatened Species until 2003, and the first endangered mushroom, northern Sicily's *Pleurotus nebrodensis*, not until 2006. A more specific census effort, the Global Fungal Red List Initiative, was started in 2013 and led to a number of workshops to develop local research capacities. The IUCN Red List has assessed 597 species of fungi as of 2022, including familiar, nonthreatened species, such as the edible yellow boletus, or "slippery jack" (*Suillus luteus*). However, efforts have focused on fungi considered to be endangered, so that nearly half (48 percent) of the 597 species are classified as threatened under the criteria used by the organization. None have been listed as formally extinct yet, but 3 are listed as possibly extinct and 34 as critically endangered.[26]

Threats to the fungus world are the same as for other branches of life: habitat destruction and pollution, adverse effects of climate change, pressure from invasive species, and even overharvesting. One example of overexploitation is in the dwindling population of the caterpillar fungus (*Ophiocordyceps sinensis*), used in Chinese traditional medicine to fight kidney and lung diseases: it is overharvested to the point that it has been classified as threatened in China and vulnerable in the IUCN's Red List of Threatened Species.

castaneae), which disappeared along with its host, the American chestnut tree, hit in the early twentieth century by a parasitic fungus from East Asia: chestnut blight. As the tree's population collapsed, so did the moth's. Although the American chestnut has survived outside its native range, and there is hope to genetically modify it to make it blight-resistant and reintroduce it into its historical habitat, the chestnut moth is gone for good.

A Look at Plants

Insects and plants are well known to be highly interdependent. Insects foraging on flowers, in particular bees and butterflies, spread the pollen and seeds of many species. The fact that insects were little affected by civilization until recently explains in part why the diversity of the plant world has not shown signs yet of excessive damage. Plants come in second position after insects, as far as their number of species: there were 350,000 vascular plants in 2020, according to the Royal Botanic Gardens, Kew (in the United Kingdom), and the numbers grow by over 2,000 a year.[27]

New plants are discovered of all shapes and sizes, from the minuscule Peruvian violet (*Viola lilliputana*), which rises less than a couple of centimeters (an inch) above the ground, to the giant dragon tree (*Dracaena kaweesakii*), a dozen meters tall (39 ft.), which boasts hundreds of branches, sword-shaped leaves, and cream-colored flowers with orange stamen. Dragon trees grow on limestone hills above the Thai forest line, but one also finds them in the courtyards of many Buddhist temples, and even as ornamental trees in private gardens, since they are reputed to bring good luck. It wasn't before 2014, however, that the species was officially recognized, proof that even a large, spectacular plant can long escape classification.

Other plants escaped detection for good reason, such as *Bulbophyllum nocturnum*: in the deep forests of New Guinea, this orchid species blooms only at night, and only one individual has been spotted so far. The plant is particularly threatened, since deforestation is progressing by leaps and bounds across its habitat.

While the census of plants is progressing fast, their extinction record has not yet reached an alarming level: approximately 600 plant extinctions have been recorded so far.[28]

One main reason, which applies to the biosphere as a whole, is that it is easier and much more gratifying to discover new species than to search for extinctions. To discover a new species, it suffices to trip upon one specimen during a survey or by accident. To prove an extinction, on the other hand, one has to comb through the plant's entire habitat, often several years in a row, until it is near certain that no individual has survived. The task is particularly unrewarding, in that there is no reward for proving a species extinct, whereas one at least gets to name a newly discovered species. Moreover, by signaling an extinction, one always stands the risk of being proven wrong: someone else can spot an overlooked survivor, which is a lucky outcome for the species but vexing for the researcher who declared it extinct in the first place.

That being said, what conclusions can be drawn from the ongoing census of plant extinctions? As was the case for insects, the list of casualties reflects more the level of effort undertaken in various parts of the world than a true assessment of the situation. Plants might also be more resilient than once thought, which does not mean that they are not at risk; simply, they will take longer to *completely* disappear than animals, and many might be quivering on the verge of extinction, without yet crowding the list of casualties. In an editorial titled "Plant Extinctions Take Time," botanist Quentin Cronk cited a case study of the Saint Helena olive tree (*Nesiota elliptica*), a buckthorn species that fell to an unsustainable population of less than a dozen individuals by the year 1900. The tree was endemic to the South Atlantic island of Saint Helena, an island overrun by goats since its discovery by Portuguese sailors in the early 1500s. Although the Saint Helena olive tree was already doomed by 1900, it wasn't until 1994 that the last plant died in the wild, and it was 2003 when the last cultivated seedling died "in captivity." Hence, while it takes 20 years for a declining animal to be assessed as officially extinct, a dying plant species can linger on for over a century before the last individual disappears.[29]

While it is important to draw a list of recent victims, it is crucial to focus on the ongoing trends of population collapse among plants and animals in order to truly assess the situation. Gruesomely called the "living dead," numerous species are headed for extinction and constitute an extinction debt that will be counted sooner or later: a time lag that masks the true severity of the situation.

6 The Plight of Endangered Species

Extinctions that were officially recorded in recent years are simply a baseline: minimum figures that are underestimates, because many animals and plants disappeared without notice, and because others are doomed—the "living dead"—but cannot be counted yet as being extinct. If we ignore these caveats and stick only to the official numbers, the status of the biosphere might appear reassuring today: extinctions at around 2 percent among birds and mammals (not taking into account the "ghost" bird species of oceanic islands, discussed in the previous chapter), less than 0.5 percent for reptiles and amphibians, and even less for fish, mollusks, crustaceans, insects, and plants. We would then be far from a great mass extinction (over 50 percent loss of species) or even from a run-of-the-mill mass extinction (25 percent loss of species). The end-Cretaceous asteroid great mass extinction wiped out 75 percent of species worldwide, and long before that, the end-Permian climate upheaval set the gruesome record of causing the extinction of over 90 percent of species.

Many experts believe, however, that what we have tallied so far, regarding our impact on the biosphere, is only the tip of the iceberg, and that extinction counts are about to increase significantly. The situation is such that, thankfully, our curiosity and sense of responsibility have kicked in, and scientific institutions all over the world are joining forces to assess the health of the bio-system and monitor as many species as possible, surveying their populations and determining their level of vulnerability and risk of extinction in the near future.

As early as 1947, UNESCO (the United Nations Educational, Scientific and Cultural Organization) piloted the creation of an international organization, bringing together natural history museums and nongovernmental organizations (NGOs) across the world: the International Union for Conservation of Nature, best known under its acronym, IUCN. In its wake was created the World Wildlife Fund (now known as the World Wide Fund for Nature outside the United States and Canada), or WWF, in 1961. Both collaborate to issue at regular intervals a Red List of Threatened Species.

A first draft of the list, established in 1949, only mentions fourteen species of mammals and thirteen species of birds in serious danger of extinction. In 1964, the official list added only one mammal to the record, but it tripled the number of birds in trouble and introduced two amphibian, three reptile, and six fish species. Fifty years later, with the help of a growing number of experts and the new tools provided by internet and computer science, the Red List has truly exploded, signaling that *tens of thousands* of species are currently threatened with extinction.

Two main criteria are used to judge if a species is threatened with extinction. One is the direct assessment of its population: Is it declining, and if so, at what speed? The other is the reduction, or fragmentation, of a species' habitat, which often precedes the collapse of its population.

Elephants: An Example of Population Collapse

Populations of many wild animals are noticeably declining. Anglers experience less catches, bird lovers count fewer birds, and car drivers observe less insects crashing on their windshields. Before jumping to conclusions, one must keep in mind that most animal populations fluctuate naturally over time, or migrate in response to weather patterns, and that a drop in numbers does not necessarily indicate a permanent downward trend. That said, the situation is truly alarming if we are to believe an estimate published in 2014 by biologist Rodolfo Dirzo and coauthors, which concludes that across the entire realm of terrestrial vertebrate animals—mammals, birds, amphibians, and reptiles—populations had dropped by 25 percent on average since 1970.[1]

The WWF periodically issues its own estimate, the *Living Planet Report*, which claims a vertebrate population (terrestrial, freshwater, and marine) drop of 69 percent between 1970 and 2018, although this higher figure is an aggregate total that mixes species with very different population sizes (mice mixed with elephants), which is somewhat misleading.[2]

One way to assess the situation is to focus on individual, representative species that best reflect overall trends and the issues at hand. African elephants constitute such an emblematic species—there are two living species: the African bush elephant (*Loxodonta africana*) and the African forest elephant (*Loxodonta cyclotis*). At the start of modern records (in the 1800s, when European game hunters began assessing their numbers), there were about twenty-five million elephants in Africa. At the beginning of the twentieth century, the population had dropped to ten million. The situation got worse with the expansion of the ivory trade, lowering the elephant population to six hundred thousand individuals by the late 1980s.

Then came a glimpse of hope. Governments acted to stop the massacre; in particular, the United States declared a ban on the ivory trade. The elephant population stabilized for a couple of decades, but hopes were dashed by the reopening of the ivory trade to many countries in 2006, followed by an upsurge of poaching and ivory smuggling: twenty-five thousand bush elephants were shot in 2011 alone, and the next couple of years were equally devastating, as poachers turned to grenades and assault rifles.

An aerial survey, dubbed the Great Elephant Census, was conducted over three years, and the results were published in 2016: the remaining elephant population in Africa was estimated at 415,000, with a yearly decline reaching 8 percent during the observation period.[3]

As a keystone species, Africa's forest elephant (*Loxodonta cyclotis*) helps maintain the structure and variety of the Guinean forests of West Africa, but it is critically endangered by poaching and habitat destruction.

The good news is that since the publication of the census, poaching has dropped to ten thousand casualties per year—mostly in response to the dismantling of gangs—which is closer to a 3 percent yearly decline. Besides poaching, however, African elephants are also threatened by habitat loss and conflicts with farmers, and despite measures taken to protect them, they qualify as "endangered" on IUCN's Red List, and even "critically endangered" in the case of the forest elephant—any change in conditions could unfavorably tip the balance.

The Plight of the Cod

The oceans are also greatly affected by population drops. Many experts believe these brutal declines signal an impending wave of extinctions—in an environment that had so far escaped massacres comparable to those on land, mostly because industrial fishing required more sophisticated and organized means. Apart from whaling, it wasn't until the twentieth century that marine species

Keystone Species

Some animals or plants play a more important role than average in servicing the ecosystem by contributing to the life cycle of many other species in their surroundings. The bee is such a keystone species, responsible for the pollination of many plants. Large animals can be keystone species as well, such as the elephant and the hippopotamus, which shape and contribute to their environment in several ways.

In the same way that the Ice Age megafauna trampled trees and shrubs, resulting in them being replaced with a grassy savanna, the bush elephant keeps the landscape in Africa open today, making it suitable for many other herbivorous species, such as antelopes and zebras. This crucial function becomes most evident when a keystone species disappears, as happened in the eastern province of Natal, in South Africa, where the bush elephant was locally eradicated at the beginning of the twentieth century. Since then, the savanna has become overgrown with shrubs and trees, causing the regional disappearance of gazelles and wildebeests as well.

Compared to the bush elephant, the forest elephant shapes the environment in a very different way. As avid fruit eaters, forest elephants distribute seeds in their dung across large distances and therefore help to replant trees. They also shape the forest by trampling across it and producing trails used by smaller animals, which also serve as firewalls, cutting off the propagation of forest fires.

As for the hippopotamus, after feeding on land, this large freshwater mammal contributes nutrient-rich dung to the waterways; the dung is a source of important elements (such as silicon), supports microbiota, and produces nutrients that flow downstream and across the food chain to benefit fish, birds, and even *Homo sapiens* fisherpeople. Hippopotami also trample freshwater bodies to create channels used by other animals, as elephants do on land, which help to flush out excess water in the case of flooding. In Africa, the decline of their population is now thought to threaten many rivers and lakes.

Many predators are also keystone species. These species preferentially hunt down weak animals and keep the stock healthy, and they keep down the numbers of herbivores in general, limiting their pressure on the environment. Since the reintroduction in the 1990s of gray wolves in Yellowstone National Park, for example, the elk population has been kept in check, and as a result, young willow and aspen shoots are less grazed, and more vigorous tree growth has brought back songbirds, beavers, and waterfowl.

Keystone species play such an important role in the ecosystem that their populations are closely monitored. Experts even created a new type of assessment for them, declaring them "ecologically extinct" when their population drops so low that they no longer play a significant ecological role, even if they are not yet threatened as a species.

became hunted on an industrial scale. The lag is being caught up today, and fish populations have joined the downward trend, as illustrated by the unfortunate example of the North Atlantic cod.

Cod was for many centuries an important source of protein and vitamins for both North Americans and Europeans, the species (*Gadus morhua*) residing on both sides of the North Atlantic, with particular concentrations on shallow banks around Newfoundland and Iceland. With the demand for cod increasing after World War II, new fishing techniques, such as purse seines (nets) and sonar tracking, drove a surge in catches that reached close to 1 million metric tons of fish per year by the end of the 1960s.

Catches then began to decline significantly, a trend first reported by coastal fisherpeople. In 1992, when it was realized that the northern cod population had collapsed to 1 percent of its historic base level, the Canadian government decided to implement a moratorium and banned cod fishing until the population rebounded. The ban was partially lifted in 1997—more a political move than a true sign of cod recovery—and turned into a catch quota of approximately 10,000 metric tons of cod per year.

This figure should be compared to an estimated stock, in 2023, of 31,000 metric tons of cod present in the species' principal habitat (the southern half of the Gulf of Saint Lawrence and Newfoundland), which translates to approximately three million individuals, if one assigns an average weight of 10 kilograms (22 lb.) to each fish. Since the cod stock was estimated at three billion individuals in the early 1960s, it has therefore dropped a thousandfold in the course of half a century, now only reaching 0.1 percent of its former level.

The bottom line, so to speak, is that the cod population is not rebounding, despite a moratorium and quotas, again illustrating how delicate balances can be destroyed within an ecosystem, driving a species toward extinction, although cod is not yet labeled an endangered species—but is at this point a "vulnerable" species of "special concern." One possible reason why the cod population has not recovered is that overfishing is also destroying the cod's prey, namely capelin and shrimp. Another reason is that climate change—specifically warming surface waters—increases

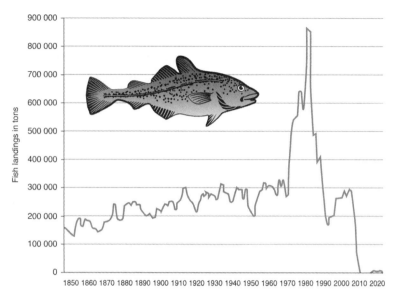

With the development of industrial fishing in the 1950s, cod catches skyrocketed off the coast of Newfoundland until overfishing led to a collapse of the fish stock in the 1970s, followed by a second collapse in the early 1990s, which lead to a fishing ban. Since the moratorium, however, cod numbers have not recovered. Adapted from Vital Water Graphics, UNEP, with data from Millennium Ecosystems Assessment.

cod mortality and allows warm-water fish, such as silver hake, to move into the cod's habitat and compete for food.[4]

Habitat Destruction: Tigers at Risk

Population collapse, whether due to overfishing or overhunting, is an obvious, direct threat for a species—a threat that can drive a species to extinction. Another critical factor is habitat destruction or fragmentation, driven by civilization's need to open up new agricultural land, quarry mineral resources, accommodate sprawling urban areas, and establish new roads and highways.

A species' habitat comprises its overall geographical range—known as its extent of occurrence—and, within it, the zone or zones where the species is effectively concentrated, known as its area(s) of occupancy. Habitat reduction can take different forms: global contraction of the areas of occupancy, impinged upon by agricultural or urban development on its borders, or fragmentation of the areas into multiple subzones, splitting an animal population into separate populations through the establishment of unsurmountable barriers, such as urban and agricultural tracts, roads, railways, or canals. Once trapped in these smaller subzones, members of a species can simply lack living space if their feeding or reproduction habits require a minimum range that is no longer provided, even if the species' *total* area of occupancy appears not to have shrunk. Habitat destruction is often a combination of both factors: a global reduction of a species' areas of occupancy *and* a fragmentation of the areas into separate parts.

One emblematic animal threatened by habitat loss is Asia's tiger, *Panthera tigris*. Poachers hunt tigers for their colorful fur and for their bones, which are used in traditional medicine, but the main problem is that the tiger's territory is encroached upon by the expansion of human civilization. There used to be nine subspecies of tigers in Asia. Two have already gone extinct in islands cut off from the continent: Bali's tiger, last reported in 1939, and Java's tiger, last seen in 1979. On the mainland, two other subspecies have fallen prey to hunting and shrinking of their habitat: the South China tiger, last seen in the late 1980s, and the Caspian tiger, which used to roam from the Caucasus mountains in the west to Afghanistan in the east and was last spotted in the mountains of Tajikistan in 1998 and declared extinct in 2003.

Only five subspecies are left, and all are threatened with extinction: the Malayan tiger, with a declining population estimated at less than 150 mature individuals as of 2022; the insular Sumatran tiger—the smallest of all living tigers—which fares somewhat better with a remaining population of less than 400 individuals; the Indochinese tiger, already extinct in Cambodia, Laos, and Vietnam, and down to about 200 individuals in Thailand, its last stronghold; and the Siberian tiger, badly off at the turn of the twenty-first century but slowly rebounding due to intensive conservation efforts, with a population estimated at close to 750 in 2022, according to Russian authorities.

The final subspecies, but not the least, is the Bengal tiger, which inhabits India, Bangladesh, Nepal, Bhutan, and southern Tibet: it accounts for over 70 percent of the total tiger population

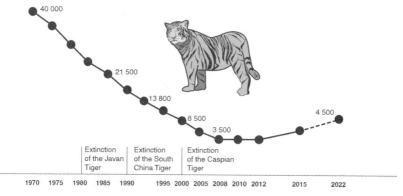

	Extinction of the Javan Tiger	Extinction of the South China Tiger	Extinction of the Caspian Tiger		

40 000

21 500

13 800

8 500

3 500

4 500

1970　1975　1980　1985　1990　　　1995　2000　2005　2008　2010　2012　　　2015　　　2022

The decline of tiger populations has included the extinction of several subspecies. The Bengal tiger accounts for over 70 percent of the total remaining population: conservation efforts in the 2010s helped stabilize the stock, and the number of tigers is slowly on the rise again. Adapted from Rodolfo Carlos Pazos / Today Online.

Siberian tigers, once on the brink of extinction, are rebounding both in the wild and in captivity. On the IUCN Red List, the subspecies has moved from critically endangered to endangered. Photo by Mathias Appel.

today. It too declined historically to alarming numbers—fewer than 1,500 individuals in the wild in 2006—but in the course of only 10 years, robust conservation efforts in India, including regeneration of forests and the establishment of reserves, have brought the population back to a number somewhere between 2,600 and 3,350 individuals in 2018, and climbing.

All subspecies taken together, the worldwide tiger population totals close to 5,000 individuals in the wild in 2022, to which one should add another 8,000 living in captivity, mostly in zoos. The population has stabilized, thanks to the efforts undertaken in several countries, but tigers remain a threatened species, especially when one considers that they numbered over 100,000 individuals in 1990: a population collapse of 95 percent in 25 years. Over the same period, the tiger's habitat shrank by 50 percent. If one excludes losses due to poaching, it would appear that the relationship between habitat loss and population drop is more exponential than linear: there are thresholds beyond which habitat loss leads to an accelerated population collapse.

With respect to tigers, one such threshold is the number of preys needed for their survival: for each tiger, about fifty boars or deer in the course of a year, equivalent to roughly one prey per week. This implies a significant roaming area, estimated at 100 square kilometers (40 sq. mi.)—nearly twice the area of Manhattan—per tiger, a challenge compounded by the fact that populations of prey are equally declining and a tiger's roaming area is bound to increase in response. Reproduction is another bottleneck: in order for it to occur efficiently, it requires a population of about a dozen individuals, male and female, that can come in contact, which corresponds in turn to a continuous roaming area closer to 1,000 square kilometers (400 sq. mi.).

The fact that civilization increasingly impinges on the habitat of tigers also leads to more frequent encounters with human beings and more accidental tiger killings. For all these reasons, and despite valiant efforts, it does not seem likely that the species will rebound to much greater numbers in the near future, leaving the small, isolated populations of tigers particularly vulnerable to any new threat that might present itself.

The Amazon Rainforest

Habitat destruction can also occur on an exceptionally large scale, as with the Amazon rainforest, the largest forest in the world. The Amazon rainforest covers an area of 5 million square kilometers

(2 million sq. mi.)—half of the planet's total rainforest area—and spreads across Brazil, Peru, and Colombia, with outreaches in Bolivia, Ecuador, French Guiana, Guyana, Suriname, and Venezuela. It harbors a rich diversity of animals and plants, which account for over 10 percent of the entire number of species on Earth. For example, the Amazon rainforest harbors 427 different species of mammals, a third of them endemic to the region—that is, found nowhere else on Earth: a full menagerie of monkeys, tapirs, armadillos, ocelots, jaguars, rodents, and bats, and in the Amazon waters, two dolphin species, one manatee, and one giant otter. The Amazon rainforest also harbors 428 amphibian and 378 reptile species (identified so far), over 1,300 species of birds, at least 16,000 species of trees—along with tens of thousands of other plant species—and hundreds of thousands of insect species, with an estimated 1 million left to be discovered.[5]

The Amazon rainforest was long protected by the fact that it was hard to penetrate. This changed when major roadways were established, beginning with the Belém-Brasilia Highway in 1958, which was followed by the Trans-Amazonian Highway in 1972. Settlers spread out along these main roadways, cutting down trees for wood and setting fire in many places to clear the forest for pasture or to plant profitable crops, like soybean. In some cases, it was simply to raise the price of land, with deforested lots being worth up to ten times the price of forested areas.

All these factors contributed to a deforestation rate of about 10,000 square kilometers (4,000 sq. mi.) a year in the early 1990s, a rate that doubled by the turn of the century and reached a record 28,000 square kilometers (10,800 sq. mi.) in 2004, comparable to an area the size of Haiti. At such a pace, the Amazon rainforest is projected to lose half its area by the end of the twenty-first century, unless strong legal and economic measures are taken to curb deforestation. Initial efforts succeeded, and forest loss was brought down to under 5,000 square kilometers (2,000 sq. mi.) in 2012, less than a fifth of what it was eight years prior, but deforestation rates picked up thereafter—notably in Brazil under the Bolsonaro government—and have been back to 10,000 square kilometers (3,900 sq. mi.) a year since 2019. These rates show no signs of abatement, despite international pressure to curb illegal deforestation.

In order for deforestation to slow down significantly, solutions must be mainly economic and rely on the perception that the Amazon rainforest brings added value to the land, more than what cleared land provides: a strategy presented in chapter 8. Another consideration is the danger of climate degradation caused by deforestation on a regional scale, since the loss of trees brings down the moisture level and causes droughts, which harm the very crops for which trees were destroyed in the first place.

While the destruction of the Amazon rainforest is relatively simple to gauge — approximately 800,000 square kilometers (almost 310,000 sq. mi.) since the early 1970s, (i.e., a loss of 17 percent in half a century) — the percentage of biodiversity loss is much more difficult to assess. A species can only be determined to have gone extinct if it was first observed alive, and at the rate at which the rainforest is being destroyed, many species are vanishing before they can even be cataloged. For a species on record, there is a time delay before it can officially be called extinct: the species must be recognized as missing through its entire area of occurrence for several years in a row.

There are also those that have not yet succumbed but are on death row and, short of a miracle, have a life expectancy of only a few years, or a few decades at most. In essence, they are temporarily surviving but have no chance of recovery. Specialists call such species, which are alive but doomed in the short run, our "extinction debt."[6]

Considering the many limitations and time delays involved in keeping tabs on the number of recorded species, experts turn to theoretical estimates, based on case studies and modeling, in order to better assess the situation in the Amazon rainforest. Some models attempt to mathematically link reduction of biodiversity to reduction in habitat area, predicting a biodiversity loss of 0.2 to 0.3 percent of species for every percent of forest destroyed. Considering that over 15 percent of the Amazon rainforest has disappeared since the 1970s, models therefore predict that anywhere between 3 and 5 percent of the total number of Amazonian species have gone extinct in the past 50 years, which would equate to anywhere from 500 to 800 tree species, 400 to 600 bird species, and perhaps as many as 20 mammal species. By comparison, current lists cite 20 bird and 10 mammal species that have officially gone extinct over the same time period, so that our official record of the situation might truly be a bare minimum figure, if we are to trust theoretical models.[7]

The Red List of Threatened Species

Assessing the number and identity of species gone extinct in historical times is a formidable task. Just as important, if not more, is assessing which species are threatened with extinction *today* and require our protection. To that end, a worldwide effort is coordinated by the International Union for Conservation of Nature: the Red List of Threatened Species. Public institutions, NGOs, and a great number of other associations provide the data, processed and compiled in a global inventory that is consultable online and updated every few months.[8]

The first observation that can be made is that the list of species assessed by the IUCN, which started off small, is growing at a fast pace. Obviously, not all 2 million cataloged species can be monitored, but efforts have been such that the number of assessed species has grown from 40,000 in 2006 to 140,000 in 2021, with a goal of 160,000 slated for the update in 2023. Some groups of animals and plants are monitored more closely than others. Because of the level of interest they generate, all mammal and bird species are followed, and because their groups are particularly threatened, amphibians are also nearing full coverage, followed by reptiles and gymnosperm trees (cycads and evergreens). Over half of all fish species are currently assessed, both in freshwater and marine environments; however, some areas are still underrepresented, such as the Arctic and Antarctic Oceans, and even the Indian and Pacific Oceans, as compared to the better-studied Atlantic Ocean and Mediterranean Sea. At the bottom of the list are invertebrates such as crustaceans and insects: only 4 percent of crustaceans and about 1 percent of all insects (12,500 out of 1 million species) are covered.[9]

Assessments are therefore still partial and suffer from two major biases: on the one hand, the most threatened groups are the most closely monitored; on the other hand, those groups that are the most difficult to assess are underrepresented. Progress is being made on all fronts, however, and as the list of assessed species keeps growing, the picture that emerges becomes more detailed and closer to reality.

The main purpose of the IUCN Red List is to classify assessed species as a function of their overall health and risk of extinction. Five categories are defined under this system. At the worst end of the scale — off the scale, should we say — are the assessed species that are found to be extinct. They are divided into two subcategories: those that are fully extinct (labeled EX) and those that are extinct in the wild (EW) but have surviving individuals in zoos and reserves.

Next comes the critically endangered category (CR), which includes species facing an extremely high risk of extinction in the wild. In effect, this is the biosphere's version of death row: if nothing is done to save them, species on this list are doomed. Significant efforts are needed to slow down their demise, reverse the trend, and hopefully upgrade them to the category of "simply" an endangered species (EN), for which the risk of extinction is very high but not critical. Next comes the category of vulnerable species (VU), for which the risk of extinction switches from very high to high. All three categories — critically endangered, endangered, and vulnerable — make up the broad domain of threatened species and are often treated as a group in reports and statistics.

Two categories fall on the positive side of the IUCN classification scheme: near threatened (NT), for species that are close to being vulnerable and might become so if no measures are taken to improve their situation, and least concern (LC), for species that need not be the focus (yet) of protection and conservation efforts. Finally, there are those species for which there is insufficient data to render a judgment (DD for data deficient), and there are those that have not been assessed (NE for not evaluated).

With respect to the 150,400 species assessed by the IUCN as of 2022, over 9,250 are critically endangered (6 percent), close to 16,500 are endangered (11 percent), and a comparable number are vulnerable. In other words, merging all three categories, 28 percent of all species assessed across the planet are endangered—that is, threatened of extinction at some level.

Groupwise, amphibians are the worst off (35 percent face extinction), followed by mammals (27 percent) and reptiles (18 percent), with birds faring slightly better at present (12.5 percent). Plant numbers are beginning to come in: out of 51,500 angiosperms (a.k.a. Magnoliopsida, the main group of flowering plants) assessed so far, 40 percent are deemed endangered—although the figure might be skewed by the fact that plants known to be endangered are assessed first. A similar figure of 42 percent is found for gymnosperms (conifers, cycads, and ginkgo); its most threatened group is the palm-like cycad order (Cycadopsida), with 67 percent of species threatened (225 out of 337). As for fungi, insects, and other invertebrates, they are vastly underassessed so far.

Comparing figures from one year to the next, one global observation that can be made is that numbers are firming up, rather than climbing, as assessment efforts continue: the IUCN report published in 2015, based on 80,000 species (rather than 150,400), listed 6 percent of species as critically endangered, 9 percent as endangered, and 29 percent as threatened as a whole—very similar to the 2022 report.

Critically Endangered

What does it take to land on the critically endangered list of the IUCN? What are the criteria used by experts—criteria that must remain sufficiently broad and simple enough to apply to groups as diverse as mammals, birds, and crustaceans without the need to invent new rules each time?

It should come as no surprise that the two main criteria to judge the health of a species are population size and habitat area. Lower limits are defined for each, below which a species is declared critically endangered. For animal species, there are five ways to make it on the list. The first and easiest way is for a species to claim less than fifty mature individuals (i.e., those capable of reproduction). Obviously, such a species is on the brink of extinction unless its population rebounds.

Such is the plight of the vaquita porpoise, a small cetacean endemic to the northern end of the Gulf of California. Its first official census took place in 1997, with a population estimate of over 500 individuals; 10 years later, the count had dropped to 150, and by 2018, to less than 20. The last search, in early 2022, found only 10 vaquitas. The main reason for their demise is illegal netting, aimed at catching shrimp and fish, with vaquitas as collateral damage. The only hope for the survival of the porpoise is a near-total suppression of illegal fishing throughout its habitat, a near impossible challenge.[10]

Some species barely pass the minimum population threshold but are still placed on the critical list because they are also experiencing the threat of habitat loss. Such is the case of the northern sportive lemur (*Lepilemur septentrionalis*), a small tree-dwelling primate on the northern tip of Madagascar that owes its name to its cute boxing-like stance when threatened. Surveys in 2012 and 2013 identified only 52 individuals, but a more thorough survey in 2019 found a total of 87. Since its forests are being destroyed at an alarming rate, however, this lemur is still regarded as critically endangered, as are 33 of the 107 species of lemurs in Madagascar. Lemurs one of the most threatened groups of mammals on the planet.

For larger populations (up to 250 mature individuals), a species can also be listed as critically endangered if its population is shown to decline rapidly, with a drop of at least 25 percent over 10 years. Another option ignores the absolute size of the population, which can be initially assessed as quite large, and focuses solely on its rate of decline. If the drop reaches 80 percent over 10 years (or over three generations, whichever happens faster), the species is declared critically endangered.

Such is the case of the saiga antelope (*Saiga tatarica*), which used to roam the Eurasian steppe from the Carpathian Mountains in the west to Mongolia in the east but is constrained today to Kazakhstan and part of Russia. Hunted mostly for their horns, saigas declined rapidly in the early twentieth century but still reached 2 million individuals in the 1950s. Following the breakdown of the Soviet Union, demand for the saiga's horns (used in Chinese traditional medicine) and unchecked poaching led to a second pop-

The European saiga antelope (*Saiga tatarica*) is threatened both by poaching and climate change. Photo by Andres Giljov, licensed under CC BY-SA 4.0.

ulation collapse, from 1 million individuals in the mid-1990s to less than 50,000 in 2003: a 95 percent drop in less than 10 years. This was more than enough to declare the antelope critically endangered. Efforts were undertaken to protect the animal, and it made a comeback by the mid-2010s, when its population returned to a healthy 250,000, but the upturn was short-lived.

In 2015, an infectious disease struck the population—discussed in the next chapter, when we address the added stress of climate change—and the saiga population dropped by more than half in a matter of weeks. Although the population has rebounded once more since then, thanks to the establishment of conservation areas and to a government crackdown on poachers, such fluctuations are deemed alarming enough to keep the saiga antelope on the critically endangered list.

A fourth way to classify a species as critically endangered is to take into account the loss or degradation of its habitat: a particularly useful tool in situations where it is difficult to perform population counts. Remote-sensing is often used in the process; for example, the reduction of a habitat can be monitored by analyzing satellite imagery from one year to the next. Under this habitat criterion, a species is judged critically endangered if its area of occurrence falls under 100 square kilometers (40 sq. mi.) and is threatened by further decline, fragmentation, or quality loss.

Finally, when it is difficult to apply any of the above criteria, experts can devise their own methodology to prove that a species has more than a 50 percent chance of going extinct in the next 10 years and deserves to be placed on the critically endangered list.

Mammals and Amphibians on Death Row

While 6 percent of all species are on the critically endangered list, two animal groups are of special interest to us: mammals, because we belong to the class, and amphibians, because they are particularly at risk.

Our mammal class is particularly well studied and nearly all species are assessed, although 14 percent are still data deficient and cannot be ranked in terms of vulnerability. Of the 86 percent that can be assigned to a list, 233 mammal species are considered to be critically endangered (as of 2022), which represents a 4 percent ratio, slightly less than average with respect to other groups but alarming nonetheless. Worse yet, 29 of those 233 species are tagged as possibly extinct at this time.

Some mammal groups are more threatened than others, namely primates, the order to which we belong. Among them, lemurs are worst off: 33 out of 107 species—about a third—are critically endangered (and all others, except four, are endangered or vulnerable). Lemurs are hunted for bush meat and affected by the destruction of their forest habitat for firewood and clearing land for agriculture. Also prominent on the critically endangered list are many Old World monkeys, including macaque, colobus, langur, and snub-nosed species, as well as the lesser apes known as gibbons (5 out of 20 species, with 14 other ones endangered), all three species of orangutans, and several subspecies of gorillas.

Many New World monkeys, including spider, capuchin, and tamarin species, are also critically endangered. All in all, out of 522 primates assessed (as of 2022), 88 are critically endangered: a ratio of 17 percent, way above the 4 percent average ratio for mammals.

Among other mammal groups, the critically endangered list includes close to 60 rodent species, encompassing rats, mice, two gophers, one marmot (living on Vancouver Island), and one flying squirrel, with habitat destruction and cats to blame in many cases; 24 species of bats, from Jamaica to New Zealand; 2 canine species (North America's red wolf and Chile's Darwin's fox); 9 wild cat subspecies, including the Northwest African and the Asian cheetah, the Mediterranean serval, the Balkan lynx (a national symbol in the Republic of North Macedonia), and the Malayan, Sumatran, and South China tigers; 17 marsupial species, including the golden-mantled tree-kangaroo, northern hairy-nosed wombat, and one-striped opossum; a camel, a gazelle, and the saiga antelope; 3 out of 5 species of rhinoceros; the African forest elephant; the African wild donkey; a couple of long-beaked echidna; and a couple of pangolin species. Half a dozen sea mammals top the list, including the vaquita porpoise, the likely extinct baiji freshwater dolphin, the Atlantic humpback dolphin, and the North Atlantic right whale.

Mammals are iconic and well followed but not the most threatened animal group: that is a record held by amphibians. Out of some 7,500 species assessed (as of 2022), 722 amphibians are critically endangered: close to 10 percent—more than twice the ratio among mammals.

The story of amphibians is enlightening. Recall that until the 1960s, mass extinctions were not fully recognized in the geological record, often because experts of different animal branches, from dinosaurs to plankton, worked separately and failed to view the global picture. Recently, albeit on a more restricted scale, the same phenomenon occurred in the amphibian world: experts focused on the decline of their pet species without realizing that their colleagues were facing similar trends. It wasn't until September 1989, during the first international congress devoted to amphibians, held at the university of Kent in the United Kingdom, that a disturbing number of researchers reported declines in their study populations, from Canada across the United States to Central America, the Andes, and the Amazon basin, as well as in Australia. This alarming realization led herpetologists (amphibian experts) to launch a comprehensive review of the state of their animal class, with the help of the IUCN. Conclusions reached a dozen years later showed that over the 1980–2000 period, a total of 9 amphibian species had gone extinct: twice the number of birds lost over the same period, while not even one mammal had gone extinct. More alarming yet was the fact that another 100 or so amphibian species were perhaps also extinct, which would take a number of years to confirm.

Indeed, although "only" 36 amphibians are listed as extinct in IUCN's official report in 2022, experts believe that the present toll is more on the order of 200 species, double the previous estimate. The fact that 722 amphibian species are listed as critically endangered confirms this trend and extends it through the present and into the future.

Major threats to amphibians include habitat loss, which stands out as the main problem for most species on Earth; overhunting, since many species of frogs are food stock, especially in Asia; pollution of ponds and rivers; and a new disorder—climate change and its adverse effects, examined in the next chapter.

A Dynamic Situation

Critically endangered species are only the tip of the iceberg. The two next categories of the IUCN Red List, endangered and vulnerable species, are based on similar criteria—reduction of population or habitat area—with less stringent thresholds.

While a population drop of 80 percent over 10 years defines a species as critically endangered, a drop of "only" 50 percent to 80 percent over the same interval will qualify a species as endangered, and if the drop is estimated between 30 percent and 50 percent, simply vulnerable. The same logic of relaxed thresholds applies to other criteria, such as the absolute number of mature individuals (less than 50 to qualify as critically endangered, less than 250 to be deemed endangered, and less than 1,000 to be vulnerable) or the size and fragmentation of habitat.

With respect to mammals, for example, whereas 233 species are considered critically endangered in 2022 (a 4 percent ratio, relative to the 5,973 assessed species), 550 are listed as endangered (9.2 percent) and 557 as vulnerable (9.3 percent). As already pointed out, all three categories—critically endangered, endangered, and vulnerable—can be grouped under the banner of "threatened species": the total reaches 1,340 (22.5 percent) for mammals, nearly one out of every four species.

The IUCN Red List offers a snapshot of the situation at any given time, and it is constantly updated. Another important factor is the trend: over the years, is the situation improving or getting worse? How many critically endangered species go extinct from one census to the next; how many simply remain on death row; and how many improve to the point of being "promoted" from critically endangered to only endangered?

Again, let us look at mammals. Over the course of a year, from 2020 to 2021, there were forty-eight changes of status: no critically endangered species went extinct, but no species were "taken off the hook" and reclassified as simply endangered. On the contrary, four species experienced a negative change of status: a gerbil species endemic to Armenia and Turkey (*Meriones dahli*) moved from endangered to critically endangered, as did a small titi monkey endemic to Bolivia (*Plecturocebus olallae*); worse yet, New Zealand's long-tailed bat (*Chalinolobus tuberculatus*) and Malaysia's banded leaf monkey (*Presbytis femoralis*) jumped up two categories from vulnerable to critically endangered. The other changes of status occurred between categories further down the scale of concern—vulnerable, endangered, near threatened, and least concern—but with a negative balance as well: twice as many mammal species moved toward extinction than moved toward safety. The following year, from 2021 to 2022, the trend was confirmed, with a dozen changes of status among monkeys, bats, and rats, and twice as many species moving closer to extinction than moved away from it.

Birds also show a negative overall trend, based on 134 changes

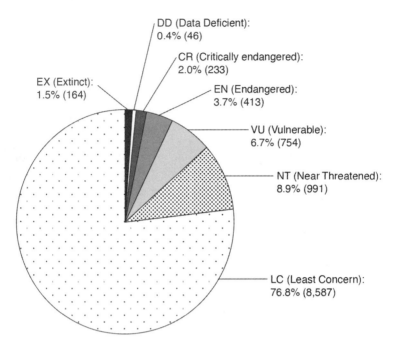

DD (Data Deficient):
0.4% (46)

CR (Critically endangered):
2.0% (233)

EX (Extinct):
1.5% (164)

EN (Endangered):
3.7% (413)

VU (Vulnerable):
6.7% (754)

NT (Near Threatened):
8.9% (991)

LC (Least Concern):
76.8% (8,587)

IUCN 2022 status of birds, indicating percentage (and number) of birds in each category; a total of 11,188 birds were assessed. Nearly a quarter of all bird species are threatened or near threatened (21.3 percent). Data from https://www.iucnredlist.org/statistics.

of status from 2020 to 2021. On the positive side, no species went extinct over the course of the year, and four got off the hook, moving from critically endangered to endangered: the Bahama oriole (*Icterus northropi*), Indonesia's black-winged starling (*Acridotheres melanopterus*), Columbia's indigo-winged parrot (*Hapalopsittaca fuertesi*), and Kenya's Taita thrush (*Turdus helleri*). In the opposite direction, however, seven species moved from endangered to critically endangered: the yellow-naped Amazon parrot (*Amazona auropalliata*), Indonesia's turkey-like maleo (*Macrocephalon maleo*), Wilkins's finch (*Nesospiza wilkinsi*), the black-naped pheasant-pigeon (*Otidiphaps insularis*), Seychelles' scops-owl (*Otus insularis*), Vietnam's crested argus (*Rheinardia ocellata*), and India's lesser florican (*Sypheotides indicus*). Again, the overall trend among birds reflects twice as many species showing a negative versus a positive change of status. The following year, in 2022, a new update of bird data led to 236 new changes of status, with this time a positive balance: three times more species improved their status than were found to be worse off. A sign of true improvement? Not necessarily. Most changes are due to better information about the reassessed species, for which the level of threat had previously been overrated, rather than reflecting a genuine improvement of their living conditions.

Reptiles fare even worse than birds, with as many as a dozen species moving from endangered to critically endangered over the course of a year (2020–2021) and another dozen jumping directly from vulnerable to critically endangered: half a dozen geckos, half a dozen skinks and lizards, ten turtles, and two tortoises. In contrast, only five reptile species moved out of the critically endangered category to a better status. The following year (2021–2022), assessments stabilized, with no extra species joining the critically endangered list and with an equal number of species improving and declining in the classification.

Because of the urgency of their situation, amphibians have been relatively well assessed over the last few years, so that over the 2020–2021 interval, the trend among their class also stabilized: half a dozen frog species declined to critically endangered, and nine frog, toad, and salamander species moved to a better status. Refreshing the data in 2022, however, indicates that the feared negative trend is quite real for the group. Although changes occur in both directions in the categories of least concern, near threatened, and vulnerable, the situation near death row is alarming: while a dozen species moved from critically endangered to endangered, eighteen species moved the opposite way, to critically endangered. Worse yet, two species went from critically endangered to extinct: the mountain mist frog (*Litoria nyakalensis*) and the sharp snouted day frog (*Taudactylus acutirostris*). Two others are feared extinct as well: the northern ticker frog (*Taudactylus rheophilus*) and the Nono robber frog (*Pristimantis hamiotae*). This is a clear example of extinction on the march.

As more data flow in for fish, a similar balance emerges, with more species being reassessed as worse off than faring better over the course of a year (2020–2021). The good news is the recovery, out of death row, of the southern bluefin tuna and the Gulf of California's totoaba drumfish. Shark and rays, on the other hand, show half a dozen newly assessed species that make their entrance directly on death row and ten more that move into the critical zone from lesser categories, including four guitar fish, four other species of ray, and two species of shark. In contrast, only one species, New Guinea's river shark, got off the hook in 2021, moving from critically endangered to simply vulnerable. From 2021 to 2022, the negative trend went on, with eight more fish species moving into the critically endangered bin—seven freshwater species and one marine (Australia's dwarf sawfish)—and, worse yet, five species going extinct, one from the critically endangered list (the Chinese paddlefish) and all others jumping several categories to extinction, as the Yangtze sturgeon did.

Long underrepresented in the census of lifeforms, flowering plants underwent 370 changes of status in 2021, more because species had been poorly assessed before then, and wrongly classified, than because they experienced a true change of situation. About as many species joined the critically endangered list (forty, including a Texan hackberry) as left it for less threatened categories (thirty-two, including one tree belonging to the yellow meranti genus of Indonesia, long threatened by coal mining and now faring slightly better). Updates continued in 2022 (another 270 or so changes): more plant species were better assessed and classified, rather than truly faring worse or better than in prior years, which again resulted in just as many species jumping up to the critically endangered list than descending from death row to a better status. One plant became listed as extinct in the wild (*Cassine koordersii*), balanced by one that was found not to be extinct (*Miconia abscondita*).[11]

The big picture is that the situation is worsening for most life groups (with the exception of plants, perhaps), which leads to the next big question: In how much time, if the trend continues, will the biosphere reach extinction figures comparable to those of the great mass extinctions of the past, such as the end-Cretaceous crisis?

Toward the Sixth Great Mass Extinction

As already noted, if we look at current numbers, one can always deny the gravity of the situation by pointing out that only 2 percent of mammal, bird, and amphibian species have officially gone extinct since the rise of human civilization, a far cry from the 50 percent figure that confers to a crisis the status of "great mass extinction," not to mention the even greater gap with the 75 and 90 percent figures reached by the end-Cretaceous and end-Permian mass extinctions. It should be recalled, however, that for many animal groups today, there is evidence that the number of historical extinctions has been grossly underestimated; for birds, as already mentioned, the loss of avian species might already exceed 10 percent.

When compared to the geological record, and even taking into account the 10 percent figure, the situation might not appear critical. The picture is different, however, if we bring into the equation the IUCN's Red List of Threatened Species: considering the number of species that are critically endangered, and the tendency over time for more species to move closer to death row than to move away from it, extinction figures are bound to increase if we stay the same course.

One can make projections based on the Red List, as biologist Anthony Barnosky and eleven coauthors did in an article published in 2011. In their theoretical model, the experts assumed that all species listed as critically endangered would go extinct in the course of the century, that those species would be replaced on death row by an equal number of species moving from the endangered category list, and that this new batch would disappear in turn over the next century, and so on. Based on this dynamic turnover model, the team projected that 75 percent of amphibians would go extinct over the next 900 years, 75 percent of mammals in the course of 1,500 years, and 75 percent of birds in approximately 2,250 years. The figures are worrisome but again do not ring a bell of urgency, since they hint that a biosphere crisis on the scale of the end-Cretaceous great mass extinction (75 percent species loss overall) would not be reached until two *millennia* into the future.

Barnosky and colleagues also performed a second calculation, in which they assumed that all threatened species—critically endangered, endangered, and vulnerable—would disappear in a century, likewise replaced by a new batch of threatened species that would go extinct the following century, and so forth. In this more drastic scenario, the eradication of 75 percent of species would occur in 240 years for amphibians, 330 years for mammals, and 540 years for birds. While our generation, our children, and our grandchildren would still not see in their lifetime the biosphere reach a level of mass extinction comparable to that of the end-Cretaceous, such a threshold would be much closer: only a dozen generations away.[12]

Obviously, this is a thought exercise that relies on assumptions impossible to assess at this time, and the authors recognize this fact. Will the species on death row—be it critically endangered species or the larger community of threatened species—effectively go extinct? Is the 100-year timeline for their "execution" well chosen, too long, or too short? Will disappearing species be replaced at the same rate by new candidates moving from safer categories? The optimistic point of view is to assume that conservation measures taken to protect endangered species will meet with success and slow down species' eradication to the point where a mass extinction will be avoided altogether.

This does not seem to be likely, however. Conservation measures have been taken for several decades, with only marginal results: for much of the biosphere, the situation has not improved significantly, at least yet, despite a few spectacular examples, such as the recovery of tigers (which will hopefully last). If one looks at the trend since 2011, when Barnosky and coauthors published their speculations, the balance of threatened species does seem to show a slight improvement, but in truth, it is only a mirage. At the time, the IUCN listed approximately 7,000 vertebrates as threatened out of 35,000 species assessed: a 20 percent ratio. Ten years later, in 2021, over 10,000 vertebrates were listed as threatened out of a larger total of 57,000 species, so that the ratio dipped slightly from 20 to 17.5 percent. Invertebrates showed a similar trend over the decade, the ratio of threatened species versus assessed species dipping from 26 to 23 percent. However, species that were known to be threatened were tallied before others, so that broadening the assessment to those other species "diluted" the numbers and caused only an illusion of improvement. In order to truly detect a trend, the IUCN recommends focusing only on those groups that have been fully assessed, and that have been so at least twice, such as mammals, birds, amphibians, and cycad plants. From one assessment to the next, changes of status among these species show a much clearer picture — that, overall, their extinction risk is worsening over time.

That extinctions are on a roll, seemingly unabated, is bad news enough. But the situation could get even worse if left unchecked. Projections, such as those published by the Barnosky team, are based on the premise that extinction mechanisms are linear: that their causes are unchanging, so that their effects on the biosphere also remain constant over time. The bigger threat, which could lead to a much-dreaded mass extinction, is for new agents of deterioration of the ecosystem to jump in and bolster those already in place, amplifying them or triggering chain reactions and leading to an accelerated breakdown of the ecosystem, well in advance of the linear model. One such aggravating factor is already at play today: global climate change.

7
Global Warming and
Chain Reactions

According to the projections made by Anthony Barnosky and his team, presented in the last chapter, our civilization is capable of creating, in a matter of a couple of millennia—and in the worst case in only a few centuries—a great mass extinction of species comparable in scale to the one that took out the dinosaurs 66 million years ago. Such speculations do not take into account any aggravating factors that might arise in the future. However, in step with industrial and agricultural growth, our civilization has induced a number of negative by-effects that stress the biosphere, starting with greenhouse warming and climate change.

The greenhouse effect relies on the capacity of various atmospheric gases to let through sunlight, heating oceans and continents, but then trap part of the infrared heat that these bodies emit back to space. Two minor gases in the Earth's atmosphere have the faculty to soak up heat in this way and raise the atmosphere's temperature: water vapor (H_2O) and carbon dioxide (CO_2). Other trace gases, present in much lesser amounts, are even more efficient at trapping heat, such as methane (CH_4)—which is two hundred times less abundant than carbon dioxide but eighty times more efficient—as well as nitrous oxide (N_2O), ozone (O_3), and various compounds combining carbon, chlorine, or fluorine, known as CFCs (chlorofluorocarbons).

Water vapor is responsible for two-thirds of greenhouse warming on Earth, but it reaches a saturation limit where it rains or snows out of the sky, limiting the amount of warming possible. Further mitigating water's greenhouse effect, liquid droplets and ice crystals form clouds, which act as cooling agents by reflecting sunlight.

Carbon dioxide, joined by a few trace gases, is responsible for the other third of greenhouse warming. Present in the Earth's mantle and brought up to the surface in volcanic eruptions, carbon dioxide has always been an important contributor to our atmosphere, as on Venus and Mars. The heating effect of carbon dioxide can be considerable if unchecked. The best example is Venus: besides being closer to the Sun, our sister planet harbors a thick atmosphere of predominantly carbon dioxide (96.5 percent), reaching 90 megapascals of pressure at the surface: ninety times the atmospheric pressure on Earth. Such a thick blanket of CO_2 traps infrared heat all around Venus, to the point where the temperature in the lower atmosphere and on the ground averages 465°C (870°F). The surface of Venus is nearly hot enough to make rocks glow at night.

The Earth avoided the fate of our "evil twin" because our atmosphere started cool enough for water vapor to condense, creating an ocean into which carbon dioxide dissolved and reacted with metals like calcium to drop to the bottom. How much carbon dioxide is left in the atmosphere is but a small fraction of the total amount; the exact proportion depends on the changing input of volcanism on one hand and the sequestration rate that scrubs it from the atmosphere on the other, which is dependent in turn on chemical transfer into rock minerals (carbonization) and its precipitation as carbon-rich ooze by plankton and other microorganisms. When conditions are right, this organic ooze is buried and compressed to form the fossil fuels—coal, gas, and oil—that our civilization exploits today.

The natural greenhouse effect, finely tuned by the interactive mechanisms at work on our planet, is a good thing: without the trace amounts of greenhouse gases present in our atmosphere, the average temperature at the surface of the Earth would hover around −18°C (0°F), and all oceans would be frozen at the surface. Thanks to the naturally occurring carbon dioxide (280 ppm; i.e., 0.028 percent, before the onset of the industrial era), the Earth's atmosphere soaked up enough solar heat for its temperature to average 15°C (60°F) today.

The biosphere itself plays a role in this delicate balance. Through unconscious, automatic mechanisms, lifeforms make constant re-adjustments and to some extent control the level of carbon dioxide to their advantage,[1] although, as we have seen, the Earth's thermostat can be reset by a variety of factors: some geological periods are warmer or colder than others. In chapter 3, we reviewed the Paleocene-Eocene heat spike, 56 million years ago, when the Earth's average temperature jumped up by 5°C (9°F) over a time interval of about 200,000 years. We also discussed ice ages, when the growth of continental glaciers and sea ice altered the Earth's thermal balance by reflecting more sunlight back to space, lowering global temperatures.

The contribution of the biosphere in regulating the Earth's climate might take strange twists, as suggested in a 2010 research letter written by biologist Felisa Smith and two coauthors. Entitled "Methane Emissions from Extinct Megafauna," the letter focused on the spectacular drop in temperature that occurred roughly from 13,000 to 12,000 years ago: the cold snap known as the Younger Dryas. The cooling coincides with a marked drop in atmospheric methane — by one-third over a few centuries — as well as with the extinction of the megafauna in North America.

The authors of the study point out that mammoths and other giant herbivores produced large amounts of methane through their digestive track (even today, our livestock contributes 20 percent of atmospheric methane). The purported massacre by hunters of the North American megafauna might explain, according to the team's calculations, a sizable proportion, if not the totality, of the drop in methane at the time. In other words, the Earth's climate could be finely tuned by flatulence.[2]

Although it might be regarded as tongue in cheek, the Younger Dryas flatulence hypothesis is elegant (so to speak) and quite convincing, even if it remains to be fully proven. It serves as a good example of bioclimatic processes at work in the Earth system, many of which remain to be studied. It also underscores the fragility of the system and acts as a reminder that when tampering with climate, our civilization is far from understanding the consequences and chain reactions it may unleash.

Humankind Responsible for Global Climate Change

The idea that humankind could accentuate Earth's greenhouse effect by pumping carbon dioxide into the atmosphere is not new. It was suggested as early as 1893 by Swedish chemist Svante Arrhenius (1859–1927), who calculated that a doubling of CO_2 in the atmosphere, mainly through industrial outgassing, would raise the temperature of the Earth's surface by 4°C (7.2°F), in full agreement with recent estimates. The prediction was not perceived as a problem at the time; on the contrary, it was seen as an advantage: scientists were pondering the past occurrence and future threat of ice ages, and so viewed industrialization and its gaseous emissions as a rampart against the attacks of Mother Nature.

The greenhouse effect of human activities became truly noticeable in the second half of the twentieth century, one reason being our demographic, industrial, and agricultural expansion—the world's population doubled between 1945 and 1975—and another being that the oceans had absorbed a large fraction of the early CO_2 surplus, preventing it from heating the atmosphere, until ocean waters became somewhat saturated and played less of a scrubbing role. It wasn't until the 1960s that atmospheric concentrations of CO_2 were measured in earnest, both at Mauna Loa, Hawaii, and in Antarctica: by the early 1980s, mean temperatures on Earth were seen to rise significantly, leading to the establishment in 1988 of the Intergovernmental Panel on Climate Change (IPCC), in charge of collecting scientific information on the issue and assessing economic impacts, future risks, and possible responses.

A number of scientists reacted to the IPCC's reports, challenging the reality of climate change, of humankind's responsibility in the matter, and of the impact it might have on civilization and the biosphere at large. Some did so with the financial support of fossil-fuel lobbies and other industrial lobbies. Analyzing these reactions to global climate change is beyond the scope of this book, but a parallel can be drawn with initial reactions to the severity and rapidity of the dinosaur-slaying end-Cretaceous mass extinction. As we saw in chapter 2, many reactions at the time were of outright disbelief, because the scholars involved were unfamiliar with the science of impacts and unprepared to accept new ideas "out of left field," with the added dimension that challenging an attention-grabbing hypothesis might place them in the limelight as well. Three decades later, in the case of climate change, it is interesting to note that one established scientist, member of the French Academy of Science, who was a key critic of the impact origin of the end-Cretaceous mass extinction (blaming it instead on drawn-out volcanic eruptions), also denied the responsibility of civilization in global climate change, blaming it on variations of solar activity (his data and deductions were erroneous as well).

In general, critics of global warming acknowledged the reality of climate change but claimed that conclusions and projections by the IPCC were exaggerated. Looking back, however, it appears that the IPCC was, on the contrary, very cautious in its analysis and conclusions. Its first report, published in 1990, acknowledged the lack of scientific certainty, suggesting, but not claiming, humankind's responsibility in the change and stating that the "rate and magnitude [of climate change] were likely to have important impacts on natural and human systems."[3]

By the second report, in 1995, data and models had improved to the point where the panel could write that "the balance of evidence suggests a discernible human influence on global climate," recognizing that it was more than 50 percent likely that human activities were to blame.[4] In 2001, the third report presented "new and stronger evidence," upping the probability of human responsibility to 66 percent.[5] In 2007, the fourth report upped the probability of our responsibility to "very likely"[6]—at least a 90 percent certitude—and in 2013, the IPCC's assessment was that of an "extremely likely" probability, over 95 percent.[7] By the sixth report, in 2021, the opening statement judged it "unequivocal that

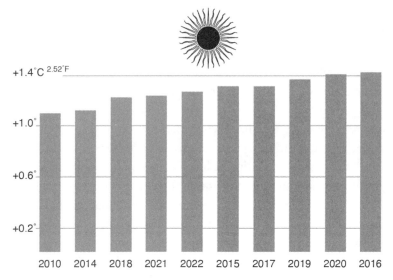

+1.4°C 2.52°F

+1.0°

+0.6°

+0.2°

2010 2014 2018 2021 2022 2015 2017 2019 2020 2016

The ten hottest years on record, expressed as their departure from twentieth-century average temperatures. Data from climatecentral.org.

human influence has warmed the atmosphere, ocean and land," and that "observed increases in greenhouse gas concentrations since around 1750 are unequivocally caused by human activities."[8]

Besides being prudent and restrained in attributing climate change to civilization, the IPCC was equally cautious in predicting temperature trends if industrial and agricultural practices, in particular the burning of fossil fuels, continue unchanged or even worsen.

The rise of global temperatures so far is well documented. By the end of the twentieth century, it had already reached 0.6°C (1.1°F) above preindustrial levels (a base level defined as the average temperature over the period 1850–1900). The rise is now observed to continue unabated into the new millennium: at the onset of the 2020s, the Earth's average temperature was about 1.1°C (2°F) higher than base level; by the year 2030, many experts believe that it will reach the high mark of 1.5°C (2.7°F). From then on, the temperature rise will depend upon how we conduct business on Earth, which the IPCC has modeled with a variety of possible scenarios. In the best possible world, where we slow down and reverse the trend of greenhouse gas emissions to values lower than today, it would still take 20 to 30 years to see global temperatures stabilize and perhaps regress, not exceeding 1.5°C (2.7°F) in 2100, but there is no sign that the world is heading in that direction.

In a more realistic "business as usual" intermediate scenario, in which emissions of carbon dioxide remain around current levels into the second half of the century, the predicted temperature rise would reach between 2°C and 3°C (3.6°F and 5.4°F) by 2100. Finally, in a scenario of sustained growth where CO_2 emissions double over the next few decades—and unfortunately the threat is real that this scenario will play out, despite the vows of many countries—the temperature reached by the end of the century would be 4°C to 5°C (7.2°F to 9°F) higher than the preindustrial baseline.

This would be comparable in scope to the temperature rise experienced during the transition from a glacial to an interglacial cycle, as happened 11,000 years ago, when the planet emerged from the last ice age; however, the new rise will build upon the peak phase of an already warm climate to bring the average temperature above anything experienced on Earth for the last *50 million years*. In other words, under these intermediate to high emission scenarios, we might be headed, in less than a century, toward a temperature spike similar to that of the Paleocene-Eocene Thermal Maximum (PETM) 55 million years ago: a crisis in its own right, discussed in chapter 3, including its marked effects on the biosphere.

A surge of several degrees will undoubtedly be damaging for human civilization, as the initial 1.1°C (2°F) rise already shows today. Besides sheer heat and forest fires, climate change with continue to intensify the water cycle, with more severe rainfall and flooding in some regions and more pronounced droughts in others.

Just as worrisome is the global sea-level rise triggered by global warming, through thermal volumetric expansion of the water mass and supplementary glacial meltwater flowing into the oceans. Starting at around 1 millimeter (0.04 in.) per year in the early 1900s, the rate of sea-level rise doubled during the last quarter of the twentieth century, and it now reaches 3 to 4 millimeters (0.12 to 0.16 in.) per year: a full 20 centimeters (8 in.) of sea-level rise over the past century. Expectations, under intermediate greenhouse and heating scenarios, are that the sea-level rise will reach 50 to 100 centimeters (1.6 to 3.3 ft.) by the year 2100, perhaps even 200 centimeters (6.6 ft.) under the most pessimistic scenarios, if ice sheets were to become particularly unstable. Flooding of coastal areas will displace and impoverish millions of people and make a deep dent into wetlands and their ecosystems.

Climate Change and the Biosphere

That civilization has to face global warming is one thing, but how will the biosphere at large react to climate change? Leaving aside the extreme 5°C (9°F) rise scenario, one might surmise that 1°C or 2°C (1.8°F–3.6°F) of global warming, triggered by moderate emission scenarios, will not have much of an effect on the biosphere, judging from comparable swings of temperature in the geological past that did not leave much trace of damage in the fossil record. Today, however, with the added pressure of overhunting, habitat reduction, and invasive species, such a temperature rise might lead to ecosystem collapse and accelerated extinctions in some cases.

Each animal or plant lives within a certain range of temperature, humidity, and other parameters: a set of suitable conditions that define its "climate envelope." In addition, each animal depends on other species—preys or symbiotic allies—that have their own vital envelopes. Also important is the timing of climate patterns, which fixes the calendar for the flowering of plants and the ripening of fruit, mating encounters and the birth and rearing of young, and departure signals of great migrations. Such seasonal cycles and signals define the phenology of a species.

In a healthy ecosystem, such cycles and timing must be finely coordinated between interdependent species. A freak weather accident can temporarily upset the ecosystem's equilibrium, but the balance is usually restored the following year. If a change of climate is rapid and irreversible, it can permanently offset cycles and unravel the complex web of interactions within the ecosystem. One of the first examples of a climate-induced breakdown took place in Costa Rica's Monteverde Cloud Forest Preserve, starting in the late 1980s, as reported by resident biologist Alan Pounds and ecologist Martha Crump in an article published in 1994.[9]

Costa Rica's Cloud Forest is a special biotope: a mountain forest, at an altitude between 1,500 and 1,800 meters (5,000 and 6,000 ft.), bathed in the mist provided by the condensation of warm tropical air. Even during the dry season, amphibians could rely on this moist haven to live and reproduce. Over a thousand golden toads (*Bufo periglenes*) were observed to gather and mate around temporary pools in the spring, until 1988 when only one toad showed up. By 1990, not one golden toad was spotted, nor were a number of harlequin frog species (*Atelopus* genus) that also used to congregate in the forest. The researchers made the connection between the amphibians' population collapse and unusual weather patterns, particularly a warm, dry spell in 1987 that lifted the altitude of the mist cloud and exposed the forest and its hosts to heat and dryness—a situation that repeated itself over the years as global warming became established.

A complication that illustrates the complexity of interplays in a threatened ecosystem is the surge of a parasite among the stressed amphibians: the chytrid fungus *Batrachochytrium dendrobatidis*, which certainly contributed to their demise. Infected frogs and toads were reported across Central America, beginning in the early 1980s. The disease is not necessarily lethal, but in a stressed environment—such as that caused by global warming—it can suddenly take on a significant role. An interesting aside is that the propagation of the fungus appears to be tied to an invasive species—the African clawed frog (*Xenopus laevis*), sold in pet stores and used in laboratories the world over—that invaded not only the Americas but Australia, New Zealand, and Europe as well.

Besides the chytrid fungus threatening amphibians, other diseases are amplified by global warming and affect a range of species, from pine trees to African lions. One spectacular example is the affliction that hit the saiga population: the critically endangered European antelope mentioned in the previous chapter. The mass die-off that occurred in 2015, when two hundred thousand saigas (over half the population) succumbed to a brutal epidemic in a matter of weeks, is believed to be a consequence of climate change. The culprit was found to be a bacterium, *Pasteurella multocida*, that is part of the animal's respiratory microbiota and normally harmless. That spring, however, the bacterium entered the saiga's bloodstream, triggering a lethal infection. Scientists blame stormy and unusually warm and humid weather in the spring of 2015 as the cause of the deadly epidemic: "a response of opportunistic microbes to changing environmental conditions," as summarized by wildlife veterinarian Richard Kock and coauthors in a thorough investigation of the epidemic.[10]

Migration as a Response to Climate Change

The negative influence of climate change on the biosphere is fostered by the fact that so many animal and plant species only thrive in their precise "climate envelope" of fitting temperature and humidity. If those conditions are altered, stress increases within the species, to the point where it can become vulnerable to other negative factors, such as pollution, invasive species, and pathogen microbes and fungi.

One would expect migration to provide an answer to the issue. If a species is unhappy, it need only find a place that satisfies its climate parameters. Currently, under global warming, isotherms (contour lines of given temperatures on Earth) are shifting poleward toward higher latitudes, as well as up mountains to higher altitudes. For a species to remain content, it must follow the temperature lines as they progressively shift across the land.

It is a fact that climate change is causing the geographical displacement of entire ecosystems today. One early and interesting study on the matter was conducted in the Green Mountain National Forest of Vermont. The progress of a species, relative to temperature, is easier to trace on a mountain slope than on flat terrain: a change of 1°C (1.8°F) occurs over an elevation distance of 150 to 200 meters (490 to 660 ft.), which corresponds to a distance on the slope of only half a kilometer (a third of a mile). Across a plain, all other weather conditions being equal, a temperature change of 1°C (1.8°F) is reached by progressing toward the pole over distances around 150 kilometers (100 mi.), depending on the region. On a mountain front, therefore, the progressive change in vegetation, as a response to global warming, is much more apparent.

The Green Mountains (which gave their name, *Les Verts Monts* in French, to the state of Vermont) were the object of a comprehensive forestry survey in 1964. By repeating the study in 2004, nearly half a century later, and relying for the interim years on satellite imagery, ecologist Brian Beckage and his team were able to estimate the amplitude and speed of change of the tree canopy in response to global warming, with the advantage that, as a protected national forest, the test area was not directly altered by human activities.

In order to follow the climate-driven displacement of the ecosystem, the chosen marker was the transition line between the downslope mild-temperature deciduous trees, such as beech and maple, and the upslope boreal trees, such as spruce and birch. In the Green Mountains, the transition normally occurs at an average altitude of 800 meters (2,600 ft.), over a range of 200 meters (660 ft.) or less. It can easily be seen, when there is no snow, as a marked change of hue: light at the bottom, dark at the top. As it followed the shift of the transition line over time, the research team established that over the 40 years from 1964 to 2004, the boreal forest, which experienced a temperature rise around 1°C (1.8°F), moved uphill by roughly 100 meters (330 ft.), replaced in its wake by the lower-slope, heat-tolerant deciduous trees.[11]

This example shows at what speed trees can be knocked out of an ecosystem, as seedlings of better-adapted species replace them. It also shows, at least in this case, that global warming is proceeding faster than the migration rate of trees. Indeed, models suggest that, at equilibrium, a 1°C (1.8°F) temperature rise should have shifted the transition line a full 200 meters (660 ft.) upslope: double what was effectively observed. Trees are therefore lagging:

in some environments, heat-threatened species might not be capable of propagating uphill fast enough to avoid being eradicated by global warming, even if there is room upslope to accommodate them.

An equivalent type of study was performed with respect to animals in California's Yosemite National Park. There, too, researchers had access to a very detailed past report, focusing on the distribution of small mammals across an altitude range of 3,000 meters (9,800 ft.), conducted between 1914 and 1920 by zoologist Joseph Grinnell.[12] Nearly one century later, evolutionary biologist Craig Moritz and his partners repeated the study, between 2003 and 2006, setting traps for small mammals along the mountain slope in order to pinpoint their location and document any change in the range of altitude that each species occupied. Like Vermont's Green Mountain National Forest, Yosemite is a National Park: human influence is negligible, and any change can safely be attributed to climate change.

Minimum temperatures in Yosemite National Park are known to have risen 3°C (5.4°F) in the course of a century: a good example of how global warming can be amplified on a regional scale, with respect to the worldwide temperature rise of 1.1°C (2°F). High-altitude mammal species with a low tolerance for heat moved further uphill to remain in their preferred cool-temperature bracket: such was the case of the alpine chipmunk, Belding's ground squirrel, and the rabbitlike pika, to name but a few. Considering available area shrinks with altitude, such an uphill migration resulted in a contraction of the species' habitat. Many mammals living at the base of the mountain range, on the other hand, gained ground: besides retaining their traditional habitat, half the documented species expanded upslope into regions now warm enough for them to occupy. The progression of mammals upslope, averaged across a dozen of representative species, was measured to reach 500 meters (1,640 ft.) over one century: a climb in altitude that corresponds to 3°C (5.4°F) of cooling. The figure perfectly matches the 3°C (5.4°F) regional temperature rise due to global warming, and it strongly suggests that the migration of Yosemite's small mammals uphill is indeed due to climate change, as expected.[13]

As previously mentioned, climbing up a mountain slope to escape excess heat creates a major problem: the higher the ground, the less area is available. There comes a point when the summit becomes extremely crowded, with no place to go: unless cornered species learn how to fly, they reach a dead end, so to speak. Migration across more level ground in response to global warming might seem less problematic, but another dead end can be reached: the shores of an ocean, blocking the poleward progression of animals and plants. In order to survive, cornered species would need to learn how to swim.

The Forced March of Species

In addition to direct human pressure, then, plants and animals must today also face the negative effects of global warming. Shifts in the makeup of the ecosystem, as those first observed in Vermont's Green Mountain National Forest and in the mammal community of Yosemite National Park, are now frequent and noticeable. A study conducted in 2003 on about one hundred species of alpine plants, butterflies, and birds of the Northern Hemisphere found an average rise in altitude of ecosystems of 6 meters (20 ft.) per decade for plants and a horizontal migration toward the north for butterflies and birds on the order of 6,000 meters (20,000 ft.) per decade. The latter figure is spectacular: 600 meters (2,000 ft.) of northern progression *each year*. Faster migrations can take place, such as happens for many bird species in Great Britain, which are moving northward at a rate of 10 kilometers (6 mi.) per decade: a full kilometer (two-thirds of a mile) per year. But in the animal world, the speed record belongs to marine plankton: many species of copepods—tiny crustaceans found at shallow depth in the oceans—shifted northward by nearly 1,000 kilometers (600 mi.) in 40 years: an average migration rate of 25 kilometers (15 mi.) per year.[14]

Migration rates in the marine world are significant because changes in water temperature occur over large distances, and strong currents allow species to move fast. On land, migration requirements are becoming increasingly difficult, if not impossible, to meet. On flat territory, species need to migrate poleward at an average rate of 500 to 1,000 meters (about a third of a mile to two-thirds of a mile) a year if they want to keep up with the latitudinal shift of temperatures caused by global warming. At first glance, such a rate appears to be within reach of most animals: even a snail can cover such a distance, which boils down to a couple of meters (6 ft.) a day.

But there is a catch. It is the whole ecosystem of an animal, including its food stock, that needs to move poleward in the face of global warming for the animal to survive. If the food stock is a plant, then the animal is held back by the plant's slow migration rate and cannot escape the heat. For plants, 1 kilometer (0.6 mi.) a year is about as fast as they can move. In fact, such a rate was achieved by trees 11,000 years ago, when the biosphere emerged out of the last glacial stage and the vegetation shifted northward.[15]

Moving any faster is a challenge, especially considering that the isotherm shift of 1 kilometer (0.6 mi.) per year is an average figure, and it is overshot in many regions. Prairie and savanna environments are particularly exposed: the rate of climate change predicted over the next decades requires that in order to remain within their preferred envelope of temperatures, flora and fauna will need to migrate poleward at a rate of 1.25 kilometer (0.8 mi.) per year, which might be very difficult, if not impossible, to meet. In fact, one study conducted in 2009 concluded that nearly 30 percent of continental land will experience poleward migration of isotherms greater than 1 kilometer (0.6 mi.) per year, which plant and animal species will not be able to follow.[16]

Moreover, these models assume unobstructed terrain, as though there were no obstacles in the race toward the poles. In such an ideal landscape, the biosphere might still be able to adapt and ride the heat wave with limited casualties, as it has done so far, each time it emerged out of a glacial stage. Today, however, the situation is different. Our civilization has added many hurdles that species need to overcome in their poleward race—fences, roads, and railroads, as well as agricultural land and urban zones—so that, in many cases, their progress is blocked. Even wildlife reserves and national parks can trap species if their borders are closed. Likewise, islands, even large ones, act as traps: such is the case of Madagascar, with its iconic baobab trees teetering on the brink of extinction (see "Baobabs in Danger").

Global warming is particularly damaging in that it amplifies the negative effects of other issues, such as habitat fragmentation. Both stress factors add up: if a habitat becomes divided into subsets barely large enough to sustain their species, the latter are very vulnerable when the added stress of global warming kicks in, since migration is impossible out of their restricted range. Hence the importance, as underscored in the following chapter, of managing "ecological corridors" between fragmented habitats in order to facilitate the roaming of animals in search of food, a mate, or a pathway poleward in response to global warming.

With respect to the latter, habitats fragmented or bound by fences are most problematic in the plains because it is on flat terrain that isotherms, and hence the climate envelopes of species, shift over the largest distance from year to year. By contrast, mountainous terrain is relatively spared. For one, as already pointed out, a slope allows species to compensate for global warming by climbing uphill, attaining 1°C (1.8°F) of cooling by covering *one hundred times* less distance than if they had to migrate poleward across the plains. Birds are sensitive to the difference: a study performed on forty passerine species in the western United States show that they migrate preferentially uphill rather than northward. The other advantage of mountainous terrain is the difficulty of access: mountain forests, for example, are less exploited than plain forests and less fragmented, which preserves large and pristine habitats.

Hence, mountain biotopes—forests or otherwise—act to a certain extent as natural wildlife reserves for the retreat of species under the stress of global warming. One major problem, however, is the heightened competition that occurs between old, established occupants and newcomers, a competition exacerbated by the shrinking of available area, as isotherms and species move uphill. Nonetheless, mountainous areas constitute, so far, a haven—a form of life insurance, as it were—for the biosphere in times of global warming.

Before the rise of civilization, mountainous areas certainly

Baobabs in Danger

Baobabs are a tree genus made up of eight different species: six are found solely on the island of Madagascar (Indian Ocean), one is known in Australia, and one in Africa and Yemen. Displaying an unusual, stubby shape, baobabs are a symbol of solidity and longevity, the oldest ones estimated to have lived 2,000 years or more, such as the Panke baobab of Zimbabwe, estimated to have been 2,450 years old when it died in 2011. The rise of global temperatures is accelerating the demise of the oldest trees and affecting the entire genus as a whole; this is combined with their shrinking habitat.

On the island of Madagascar, baobabs are indeed declining, their habitat destroyed both by urban sprawl and climate change. As early as 2013, ecologist Ghislain Vieilledent and his team established, using satellite imagery and computer models to integrate projections of what the climate will be in Madagascar over the next decades, that three baobab species are currently threatened with extinction (*Adansonia grandidieri*, *Adansonia perrieri*, and *Adansonia suarezensis*).[17]

The *Adansonia grandidieri* species counts many individuals over a large area. It does not appear particularly threatened by climate change; its main factor of stress is by far habitat reduction, as its forests are encroached upon by agricultural land. The climate envelope is bound to shrink for the two other species, however—mostly because droughts are becoming more frequent in their occupation zone.

played the same role each time the climate warmed—for example, at the onset of each interglacial stage, when the temperature surged. Mammoths, bison, elk, and other megafauna acclimated to cool weather migrated not only poleward but uphill in the mountain ranges of Europe, Asia, and the Americas. At the onset of the last interglacial stage, however, starting 11,000 years ago, human hunters chased the animals up into their sanctuaries, which acted as a dead end rather than a refuge. Today, nonetheless, mountainous areas might still serve as ecosystem fortresses, within the limits of available space.

In the biblical account of Noah's ark, the shipload of rescued animals—fleeing a sea-level rise in that particular case—ran aground at the top of a mountain (often symbolically identified as Mount Ararat in eastern Turkey), from which the preserved animals spread back into the plains. One can draw a parallel with the present situation: if civilization manages to rein in the situa-

Adansonia perrieri numbers fewer than 250 trees—a critically endangered species by definition—and its climate-tolerable habitat is projected to shrink from 21,000 square kilometers (8,100 sq. mi.), estimated in the early 2010s, to a projected 6,500 square kilometers (2,500 sq. mi.) by the year 2080: less than a third of its present area. As for *Adansonia suarezensis*, a species that currently numbers approximately 15,000 trees, the numbers look good today, but it might see its tolerable habitat shrink from 1,200 square kilometers (460 sq. mi.) to a pocket of 17 square kilometers (less than 7 sq. mi.) by the year 2080. It is already listed endangered at present.

To protect baobab trees, besides caring for the existing survivors, one strategy is to plant new trees in other favorable ecosystems, even slightly cooler than what their climate envelope warrants, in order to anticipate the global warming trend. Besides their aesthetic and cultural worth, baobabs profit from another advantage in their struggle against extinction: they are economically interesting, so that their protection will receive close attention. Baobab seeds and dried fruit pulp powder, boasting high fiber, vitamins, and antioxidants, are used in gellants, flavoring agents, and cooking in general; all uses combined, baobabs represent a market value of over $1 billion per year.

tion and conditions improve, it is again from the mountains that surviving species will descend to repopulate the planet.

Worst-Case Scenarios

Global climate change affects the entire Earth, but in some regions, instead of being progressive and manageable, it can run amok. One critical example is the Amazon basin and rainforest, damaged by both questionable management and by global warming. As mentioned in the previous chapter, clearing for agriculture and the exploitation of timber has shrunk the rainforest by 17 percent so far. Another 17 percent is severely degraded. Global warming further affects the forest biosphere in two ways: by amplifying the stress on species that are already endangered and by affecting new biotopes that were relatively spared up until now.

Deforestation first occurred in the Amazon basin on flat terrain that was easy to work and harvest; hilly and mountainous terrains were long left aside. In the coastal state of Rio de Janeiro, for instance, only 20 percent of the original forest is left standing at low elevations, less than 200 meters (660 ft.) below sea level, whereas 90 percent of the forest is untouched above 1,300 meters (4,300 ft.). In the ravaged lowland forests, a great number of bird species are endangered, some critically, because of the destruction of their habitat. The threat of global warming today is that mountain birds, which were spared until now, will be threatened in turn, as isotherms move upslope and birds need to retreat higher up into less spacious habitats.

The Amazon rainforest demonstrates another ill effect of climate change. Trees have the faculty of absorbing water vapor from the air and then releasing it back into the atmosphere (a cycle known as evapotranspiration), which can trigger rainfall. Hence, vegetation and rainfall reinforce each other, and tinkering with one can affect the other. Studies based on climate models demonstrate that the more the Amazon rainforest shrinks the less rainfall occurs in the area, which is a direct threat to the agricultural economy: crops will become less profitable than the forest land that was sacrificed for them.[18]

The scenario could well turn into a disaster. During the summer of 2005, the Amazon basin suffered the most severe drought of its history, accompanied by seventy-three thousand fire outbreaks across the forest. A drought of equal magnitude threatened the Amazon the following summer, in 2006, but luckily did not play out. Four years later, in 2010, a second major drought took place, followed by a third one in 2015–2016 that established a new temperature anomaly record and affected 13 percent of the rainforest (versus 10 percent in previous crises). Such repetition of extreme conditions constitutes a major threat to the rainforest: field experience and climate models demonstrate that three consecutive years of extreme arid conditions could trigger the collapse of more than half the Amazon rainforest.[19]

The sudden death of millions of trees during each drought has immediate consequences on the Earth system as a whole. During normal years, the Amazon rainforest scrubs 2 billion metric tons of carbon dioxide out of the atmosphere (mainly through photosynthesis), limiting the extent of global warming. However, in times of drought, as in 2005 and 2010, the Amazon basin released 4 billion and 6 billion metric tons, respectively, of CO_2 back into the atmosphere through forest fires and other drought-induced tree destruction, the equivalent each time of 20 percent of civilization's yearly CO_2 emissions (or as much as China's and Russia's total emissions combined).

The problem is all the more acute that, after a decade of efforts to curb deforestation, clearing for agriculture and mining resumed in the mid-2010s and intensified during Jair Bolsonaro's presidency (2019–2022). Scientists are concerned that the Amazon rainforest might reach a threshold beyond which negative feedback loops will kick in: the shrinking of forested area triggering droughts, which in turn accelerate forest loss in a vicious circle. Studies indicate that the repetition of extreme events is destabilizing the rainforest to the point where it is losing resilience — resilience meaning the ability of a system to maintain key functions when disturbed. In other words, the rainforest is becoming ever more fragile to future sources of stress. The ultimate threat is that the Amazon basin might turn into grassland and desert in a matter of decades, with catastrophic repercussions on the economies of the countries concerned and on the world climate as a whole.

One urban myth, on the other hand, is that forests like the Amazon generate a large part of the oxygen in our atmosphere through photosynthesis and that this function is at risk. In reality, forests merely recycle a small part of the abundant stock of oxygen in the atmosphere, and their net contribution to the stock of oxygen is close to zero. Rather, it is marine plankton and photosynthetic bacteria in seawater that provide most of our oxygen supply. Which brings up other questions: Is plankton itself in danger? Are parts of the marine ecosystem as much at risk of a global breakdown as the Amazon rainforest appears to be?

Oceans at Risk

Up until now, we have said relatively little about the state of the oceans, mostly because their large-scale exploitation has been lagging, relative to upheavals on land, with less visible effects—and also because marine species are more difficult and expensive to track. The focus is still on population collapse of marketable species due to overfishing, rather than on the threat of extinctions as a whole, but a population collapse is the logical prelude to extinction, and it makes the species vulnerable to any other source of stress.

Oceans respond actively to global warming by acting as a moderator in two ways: by absorbing a large fraction of the heat generated in the atmosphere by greenhouse warming and by absorbing part of the carbon dioxide itself. The first buffer effect—the absorption of energy—has so far served to slow down the climb of global temperature in the atmosphere, and in no small way. Currently, the oceans absorb 93 percent of the excess heat produced by civilization. Considering an extra 3 percent is absorbed by the thawing process of sea ice and land glaciers and 3 percent by soil, only 1 percent of global warming goes into the rise of atmospheric temperature. If oceans cease or slowdown their intake of energy—for instance if surface waters undergo less mixing and become saturated—the atmosphere will keep a much larger share of the excess energy.

In the meantime, although they soak up 93 percent of excess heat, oceans have warmed by little more than 1°C (1.8°F) over the past century. The reason is that oceans are two hundred times more massive than the atmosphere, and they heat up at a proportionately much slower rate. The global warming of oceans manifests itself, nonetheless, and quite visibly through the rise of sea level. Part of it is simply due to thermal expansion of the seawater as it soaks up heat: approximately 1 millimeter (0.04 in.) of sea-level rise per year. The other contributor to sea-level rise, and the largest, is the melting of land-based glaciers, which pour out billions of tons of fresh water each year into the oceans; this accounts for 2 millimeters (0.08 in.) of rise each year. These values seem ridiculously low, but they are accelerating from year to year, and if one thinks in decades rather than years, the figures become significant: sea level is expected to rise by at least 20 centimeters (8 in.) by the year 2050. By the end of the century, the outlook is even worse: even if we completely stop emitting CO_2 now, because of the inertia of the climate system, sea level will keep rising and reach 60 centimeters (2 ft.) above today's mark. If we do nothing to curb our greenhouse gas emissions, it might reach 1 or 2 meters (3–6.5 ft.).[20]

This is still without considering scenarios in which continental ice sheets, such as those in Greenland and Antarctica, fail and melt catastrophically, resulting in a much more pronounced sea-level rise. These figures are also a minimum, in that they can be locally amplified by the geography of ocean basins and regional climates, leading to more frequent and more pronounced coastal flooding in many places. Low-lying countries and provinces are affected across the globe, like the coastal islands of Bangladesh, which experience 5 millimeters (0.2 in.) of sea-level rise per year—almost double the worldwide average. Combined with the greater frequency of storms, the unrelenting rise is eroding beaches and riverbanks, and it is destroying homesteads and arable land.

As for the marine biosphere itself, it already suffers from the temperature rise, even if 1°C (1.8°F) in the oceans appears to be a modest figure. One would expect mobile species, like fish and plankton, to simply migrate toward higher latitudes, or to greater depths, in order to find cooler water. Marine species are already seen to migrate poleward at an average rate of 7 kilometers (4 mi.) per year: five times the average migration rate of terrestrial species on flat land, because differences in water temperature are spread out over much larger distances. On the other hand, not all mobile sea creatures can migrate successfully, as they may be dependent not only on the right temperature but also on the right salinity, sea-bottom texture, and nutrient supply: all conditions have to be met.

Global warming is even more stressful, of course, for sessile species: those fixed to the sea bottom, like coral reefs, which migrate much slower, like trees do on land. Coral reefs are important marine ecosystems. Coral is the association of polyp invertebrates that live in colonies and build a protecting structure and monocellular algae (named *zooxanthellae*), which live within their tissue. The algae perform photosynthesis—hence the reason why coral

Coral bleaching occurs when temperature rise or other factors cause colorful algae to die or leave their hosts, which leads to the death of the coral. Photo by Vardhan Patankar, licensed under CC BY-SA 4.0.

reefs need to be close to the surface—and produce organic nutrients for the polyps; in return, the polyps provide the structure to maintain the algae close to the surface and supply them with carbon dioxide to keep the photosynthesis going.

There are over two thousand species of coral across the world, and their reefs host rich communities of marine animals, ranging from sponges and mollusks to crustaceans and fish. All in all, an estimated one-quarter of marine species are believed to live in coral reefs: a concentration of biodiversity comparable to that of rainforests on land. Reefs also act as barriers to protect coastlines from storms and erosion.

Coral reefs, however, are fragile systems, very sensitive to the impact of human activity. They are overexploited—overfished, ripped up by boat anchors and excessive scuba diving, threatened by anarchic coastal building, and hit by pollution, and hit by turbidity, which decreases their light supply—and now they are affected by global warming as well.

Most coral species prosper in a narrow seawater temperature

range of 26°C to 27°C (79°F to 81°F), which explains why they are mostly found in equatorial and tropical zones, between 30 degrees of latitude north and south. A few species live in cooler waters, including in the Mediterranean; others have adapted to the warm water of lagoons, coastward of the reefs; but in general, coral dislikes abrupt temperature jumps: a rise of 1°C (1.8°F) lasting several months is often unbearable and leads to coral bleaching: a spectacular loss of color that indicates that the coral's pigment-rich algae have died off or left their host body. This can lead to the death of the coral.

Coral bleaching was noticed for the first time in the Caribbean in 1979; it reached global proportions during the summer of 1998—a record year of global warming—when an estimated 16 percent of all coral perished worldwide. Since then, crises recur and intensify, to the point where an estimated 75 percent of coral reefs are now threatened, due to both direct human aggression and global warming. Projections are no better: 90 percent of all coral reefs might be threatened by 2030, and the entire stock by 2050, if trends are not reversed. The threat on coral does not only take the form of reef destruction and population collapse—it also carries the risk of extinction for its constituent species. In 1998, a first assessment carried on 700 coral species found only 12 to be endangered. Much more work and 25 years later, in 2023, the figure had jumped to 231 threatened species.[21]

Coral-dependent species are also feeling the stress of the warming waters. In the wake of the European heatwave of 2003, the northwest Mediterranean coast lost millions of sponges, mollusks, and other invertebrates: an affliction tied to the development of bacteria that spread across the coral colonies.

Rising water temperature is not the only ill effect of global climate change on marine communities. An equally worrisome threat is the increasing acidity of seawater. Almost a third of the carbon dioxide released by human activities into the atmosphere dissolves into sea water: a buffering effect that limits global warming, adding to the direct absorption of energy. The flip side of the coin is that a large proportion of this dissolved gas turns into carbonic acid (H_2CO_3), raising the water's acidity.

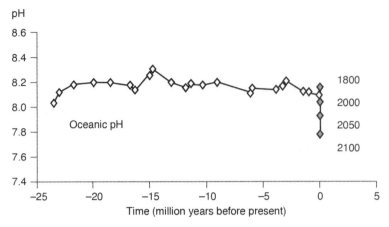

The alarming pH drop of seawater over the past century is establishing a record of acidity unparalleled in 25 million years. Data from the European Environment Agency.

The acidity or basicity of an aqueous solution—that is, its concentration of reactive hydrogen ions (H^+)—is denoted by its "pH" (which stands for "potential of hydrogen"). By definition, a pH of 7 stands for a neutral solution: neither acid nor basic. A pH smaller than 7 denotes an acidic solution, and a figure larger than 7 a basic solution. The scale is logarithmic: a change of one pH unit corresponds to a tenfold change in the concentration of hydrogen ions. So even a small change in the pH number is significant: a drop of 0.1 means the acidity has increased by 30 percent.

In essence, seawater is slightly basic. Two centuries ago, before the Industrial Revolution left its mark, surface waters registered approximately 8.15 on the pH scale. Over the years, the dissolution of CO_2 into the oceans and the concomitant production of carbonic acid has led to the progressive acidification of seawater, expressed by a constant decline of its pH number: in 2013, it was down to 8.1 in temperate waters (and 8.05 in hotter tropical waters), a trend confirmed in 2023 when the average seawater pH was found to be 8.06, corresponding to roughly a 30 percent increase in acidity. By the end of the century, the pH is predicted to dip below 8, which would represent more than a 50 percent increase in acidity over preindustrial values; under extreme global warming scenarios, it might even plummet to 7.8—double the preindustrial levels, and the highest acidity reached by the oceans for at least 20 million years.[22]

In an absolute sense, such an acidity increase might not seem like much. After all, milk has a pH of 6.5 and coffee a pH of 5, a thousand times and a million times more acidic than seawater, respectively, so how important is a mere fraction of a pH? Slight differences do matter, however. Marine lifeforms prosper within a narrow envelope of temperatures, salinity, and acidity, so that even a minute departure from their proper living conditions can lead to disfunction, population collapse, and possible extinction. The rate at which these changes occur is also important, since it determines whether species have time to adapt, through bio-chemical adjustments or mutations. The fact is that seawater acid-ification is occurring much faster today than at any other time in the past 55 million years, since the last major upset of the bio-sphere at the Paleocene-Eocene boundary (the PETM event, see chapter 3). Indeed, the rise of acidity is occurring ten times faster than it did during the PETM crisis.

The acidification of the world's oceans is particularly damag-ing for species that build their protective shells out of calcium and carbonate ions dissolved in seawater. Whenever acidity increases, carbonate ions become increasingly rare, because they combine with hydrogen ions to form salts of little use to marine organisms. Not only do free carbonate ions become scarcer as a result but those already in use within shells and skeletons start dissolving back into the waters, weakening the organisms that depend on them for protection and survival.

The destabilization of organisms dependent on calcium car-bonate has already begun. Australia's Great Barrier Reef experi-enced a 10 percent decline of its calcification rate between 1990 and 2005, according to research conducted by Glenn De'ath and partners at the Australian Institute of Marine Science.[23] Rising acidity and water temperature are expected to uphold the trend throughout the century: a tipping point could be reached by the 2050s, when coral reefs are predicted to switch from net growth to net dissolution and die off.

Most shell-producing mollusks, like oysters, are also sensitive to water acidification, especially in their larval stage, during which they need to expend much energy to combat the excess acidity of the environment, often to the detriment of other vital processes, which leads to increased mortality. In Oregon and Washington, coastal oyster farms are particularly affected. Here, it is not so much the slow acidification of surface waters that is to blame but global climate change, which triggers occasional upwelling of deep cold water—carrying more dissolved acid than surface water—and episodically swamps the fisheries. This was first observed in oyster farms on the Oregon coast, between 2005 and 2009, when the production of oyster seeds dropped by 80 percent. Acidic upwellings are now observed on the East Coast as well: if no action is taken, shellfish harvests could decline in the United States fisheries by 25 percent over the next 50 years, which equates to losses of hundreds of millions of dollars.

Plankton fares no better. In one landmark study, paleontologist K. J. Sebastian Meier and coworkers focused on *Emiliania huxleyi,* an abundant phytoplankton species that they sampled in the northwest Mediterranean between 1993 and 2005. The platelets of calcium carbonate secreted by the plankton—which resemble miniature ball bearings and are named coccoliths—displayed a weight loss of 20 percent compared to the norm, mostly due to reduced calcification. This is the lowest weight recorded for this species over the past 10,000 years, although it remains to be proven whether this thinning trend is due to acidification as such or to the broader effect of global climate change, in which the competition of other types of plankton promotes the lighter forms of *Emiliania huxleyi* through natural selection.[24]

The thinning trend appears to be worldwide, however, and not limited to one species or region in particular. A study conducted in the Southern Ocean, off the coast of Antarctica, indicates that another class of plankton species, foraminifera, experiences a similar decrease in carbonate shell weight, on the order of 30 percent. It is yet unclear how these changes will affect the health of plankton in the long run or, in turn, how this will affect the fine exchange of CO_2 and O_2 between atmosphere and ocean, which is ultimately controlled by plankton. Another concern is the relationship between water temperature and the balance of saturated and unsaturated fatty acids (such as omega-3) in the plankton makeup. Unsaturated fatty acids are essential nutrients for marine animals up the food chain, and for humans as well. Under worst-case warming scenarios, the amount of plankton-produced vital unsaturated acids might drop by 25 percent by the end of the century, and the world's food stock with it.[25]

Because plankton is at the base of the marine food chain, it is of paramount importance for us to understand such changes and trends if we wish to correctly assess the health of the marine biosphere and predict its future. One of the big unknowns, at this stage, is how fast species might adapt to offset destabilizing trends in their environment. Another big unknown—the topic of the next chapter—is to what extent human beings are capable of slowing down our aggressions on the natural world—mending the damage already caused and finding original solutions for the future.

8 How to Fight Extinctions

There is no longer any doubt that the Earth is undergoing a surge of extinctions well above the background rate. The current rate is unprecedented since the asteroid-driven end-Cretaceous mass extinction, 66 million years ago. Even if we were to magically stop extinctions today, the toll is already heavy for many groups of plants and animals.

The collapse of the megafauna, as the Earth pulled out of the last glacial stage, constitutes the first pulse of extinctions; it was followed, 10,000 years later, by a second pulse that affected oceanic islands, overrun and destabilized, first by Polynesian then by European sailors and settlers, with as many as 10 percent of all bird species exterminated in the process.

Finally, the third wave of human-age extinctions began two centuries ago with the onset of the industrial era, intensive agricultural practices, and unrestrained growth of the human population. How pronounced and lasting the ongoing wave of extinctions ends up being will depend on the efforts deployed by our civilization to contain it: namely, on reasonable scenarios of economic growth, the curbing of global warming and climate change, and dedicated programs to protect and preserve the biosphere.

At one end of the scale, there is an attitude that consists of contesting the importance of extinctions—and of climate change. Species are vanishing: so what? As long as the most profitable and useful species remain plentiful and keep market stalls well stocked, who needs the other ones? Minimizing climate change and extinctions serves the agenda of conservative policy makers, oil and coal companies, and other large corporations that conduct and promote business as usual.

Less cynical but equally paralyzing is defeatism: many people feel powerless, faced with uncommitted or slow-moving corporations and governments, and they believe that since decisions are made above them, on the national or international level, their opinions and attitudes count for nothing. Fortunately, this is not the case. Although many governments are still slow in implementing policies, there is increasing pressure from voters (at least in bona fide democracies) for governments to deliver on their promises. Citizens can make a difference, not only through their votes but also by setting an example and personally changing their living habits, in particular their energy consumption and food diet.

On an institutional level, much work has already been accomplished by scientists federated under the umbrella of international organizations, such as the IUCN and WWF, with respect to the identification and protection of endangered species.

Global warming and climate change cause even greater concern, because they directly impact the worldwide economy—floods and droughts come with a cost. This led to the creation, in 1998, of the International Panel on Climate Change (IPCC), under the auspices of the United Nations. Every six to seven years, on average, the panel issues a comprehensive assessment report, pooling the results of published research, including projections of future climate, as a function of greenhouse gas emission scenarios. The reports provide a solid database to define policies, an effort undertaken by the United Nations Framework Convention on Climate Change, which brings together in a yearly conference close to two hundred member states (known as the Conference of the Parties, or COP). One major accomplishment of the 2009 session (COP15, held in Copenhagen) was the signing of a commitment by nearly all countries to limit global warming to less than 2°C (3.6°F) over preindustrial values. Further conferences, such as COP21, held in Paris in 2015, and COP26, held in Glasgow in

2021, were the stage of enhanced commitments to mitigate global warming, such as explicitly reducing the use of coal and achieving zero greenhouse gas emissions by the second half of the century.

Commitments and strategies to reduce global warming demonstrate that when the economy is at risk, actions can be taken, or at least discussed. The preservation of species is another challenge altogether, but it shares a number of issues, problems, and solutions with the international battle against global warming.

The Root of the Problem

Both global climate change and the extinction of species share a common root: our very lifestyle, which is based on an unbridled consumption of energy and raw materials, in step with an ever-increasing world population. Up until the second half of the twentieth century, the expansion of the world population and of its material needs met with few obstacles or limitations: the planet's resources were believed to be infinite, and epidemics and warfare served to keep the human population in check. Medical advances and the relative stability of industrial states in the wake of World War II, however, resulted in an economic boom that revealed the limitations of the consumer society, exposing its main caveat: economic growth does not take place in an open, boundless system but on a "closed" planet with finite resources. From this awareness stemmed the alternate strategy of sustainable development, which promotes the recycling and renewal of resources—preserving the stability of the ecosystem in particular—but still within the framework of economic growth. More radical yet is the concept of negative economic growth, which challenges the very foundations of modern civilization by calling for restraint in material consumption, which should be limited to essential needs only.

Both sustainable development and negative growth hinge on a major reconsideration of our use and sources of energy. Fossil fuels, such as coal, gas, and oil, are our main sources of carbon dioxide—hence of global warming—so that switching to nonpolluting, renewable energies, such as hydroelectricity, solar energy, and wind power, has become a priority, and an urgent one. The

IPCC has constantly upped its warnings over the years, finally stating in its 2022 report that if insufficient action is taken *now*, the well-being of civilization and of biodiversity risks irreversible losses.

Along with global warming and pollution, the key factors that threaten the biosphere, as discussed in previous chapters, are overexploitation of species, mostly for food, and habitat destruction. Some of the most urgent measures to uphold and improve, then, are to strictly control hunting and fishing through quotas and to fight poachers, as well as protect habitats through parks and other natural and artificial reserves.

Simple in theory, such measures are often difficult to implement. Recognizing that a species is endangered, for instance, usually comes too late for quotas or moratoriums to remedy the situation, especially if the species' population dips below a threshold where negative feedback loops accentuate its downfall, as exemplified by the collapse of the cod population. Planning ahead is also particularly difficult when hunger prevails among human populations. Rare and endangered species, which were once spared, become bushmeat, as many lemurs in Madagascar and a number of monkey species elsewhere have. There are no simple solutions, then, especially when the countries involved collapse politically, along with their regulatory bodies. In such situations, one possibility is for more prosperous countries and nongovernmental organizations to provide financial support to maintain preservation efforts and assist educational projects. In an open letter published in 2010, Julia Marton-Lefèvre, director general of the IUCN, urged the thirty member states of the Organization for Economic Co-operation and Development (OECD) to commit, in addition to the 0.5 percent of their gross domestic product earmarked for economic assistance, 0.2 percent for the conservation of biodiversity within developing countries.[1]

One focus of such programs is teaching local populations how to reasonably utilize species without driving their numbers down or, in some cases, without consuming them at all. As an example, lemurs in Madagascar represent a precious tourist attraction: showcasing the animals is much more profitable in the long run than killing them or felling their trees, the caveat being that the income generated must go to the people on the spot who need it most, and not to some intermediary or bureaucratic caste.

Organized poaching is another major concern. In Kenya, the illegal butchering of elephants and rhinoceroses for their tusks and horns reached critical levels in the 1980s, and after a respite due to an international ban on the ivory trade and the establishment of a well-organized Kenyan Wildlife Service, poaching took off again at the turn of the century, partially in response to economic growth in Asia driving increased demand for horns and tusks. The situation is improving again, through reinforced cooperation across borders—between Kenya, Uganda, and Tanzania in this case—and thanks to the support of international police organizations, such as Interpol, in the field and at airports and seaports. In addition, renewed efforts are aimed at discouraging and educating the buyers themselves by debunking traditional beliefs as to the healing virtues of horns and tusks.

In response to these efforts, the year 2020 marked a turning point, when not a single rhinoceros in Kenya was lost to poaching; after dipping below 1,500 individuals the previous year, the population of rhinos rebounded past the 1,600 mark. Likewise, while 350 elephants were killed illegally in 2015, only 11 were declared lost in 2020 (but this was during the COVID-19 pandemic, when population movements, including those of poachers, were severely curtailed). These are also minimum figures, since many killed elephants are not recorded, but in any event, it appears that elephant poaching in Kenya has dropped to less than 100 elephants a year. Without slowing down its war against poaching and organized crime, the country can now focus on other threats, such as global warming: during the first half of 2022, in Kenya's Tsavo National Park, at least 179 elephants died of thirst, versus fewer than 10 officially lost to poaching, which led Najib Balala, Kenyan Wildlife and Tourism minister, to declare to the BBC: "We have forgotten to invest into biodiversity and management and ecosystems. We have invested only in [fighting] wildlife trade and poaching."[2] As well, since 2017, Kenya has been suffering its worst drought in 40 years; one solution is for the country to plant

Bollywood actress Nargis Fakhri stands beside the world's last remaining male northern white rhino, at Ol Pejeta Conservancy (Kenya) in 2015, to raise awareness and support for endangered species. The rhinoceros died of old age in 2018, causing the subspecies (*Ceratotherium simum cottoni*) to be now extinct in the wild. Photo by Make it Kenya/Stuart Price.

drought-resistant trees to restore forests, shade, and water capture.[3]

In terms of poaching, Kenya's success story shows what can be done to protect endangered species. Other countries might be less successful if their governments are less committed or politically unstable. The salvation of a species can then be helped by making sure it can disperse between neighboring countries, across borders, which multiplies its chances of finding at least one country that offers the right conditions for survival.

Natural selection can also play a role in circumventing the

threat posed by human beings. In a study published in 2021, reflecting on the impact of ivory poaching in Mozambique during 15 years of civil war (1977–1992), evolutionary biologist Shane Campbell-Staton and coauthors noted that among surviving elephants, the number of females that were born tuskless—a genetic possibility in the DNA lottery—jumped from 20 percent before to 50 percent. Tuskless elephants were not killed, since they had no market value, and thus offered a natural survival solution to the species.[4]

Efforts to protect specific species can meet with success, as shown by elephants in Kenya and tigers in India. But if the threat of a human-caused mass extinction is to be avoided, such efforts need to be sustained globally and permanently, which is far from being the case right now. In a report published in 2010 on the status of vertebrates around the world, Michael Hoffman, former chair of the IUCN Red List Committee, and his 173 coauthors estimated how many threatened species were recovering in response to measures taken to protect them. They did this by counting which species had changed categories in the Red List, by improving their status over the time interval examined—up to 20 years (1988–2008), depending on the data sets available. For mammal species threatened by overhunting and poaching, for instance, only six experienced an improvement in status, whereas sixty-two progressed negatively, toward extinction, despite protective efforts. Protective efforts have been more successful with birds, since shifts in the Red List show nine bird species improving their status, while thirty-one experienced a decline over the same period. Although the net balance is negative for all classes of animals, there is a glimmer of hope in that the authors determined that without conservation efforts, the deterioration trend would have affected at least 20 percent more species.[5]

Protecting Habitats

Habitat loss is one of the most difficult threats to combat. In the above study, the authors estimated that "for every 10 species deteriorating as a result of agricultural expansion, fewer than 1 improved because of mitigation of this threat." Mitigation usually consists of establishing reserves and other protected locations. Their total area of these locations is constantly rising and covers 17 percent of all land on Earth and 8 percent of the world's oceans as of 2022.

Such figures are encouraging but require some prudence. In particular, total areas can be misleading: in order to be efficient, reserves must be individually large, not scattered in small pockets, to avoid habitat fragmentation and its disruptive effects. Reserves must also be strategically thought out to encompass a variety of threatened species and take into account future trends, such as global warming; they must also be carefully managed, since reserves are of little use if they are insufficiently monitored and protected.

All in all, conservation efforts are beginning to pay off. To monitor the recovery of species, researchers at the IUCN developed a new method of analysis, named the "Green Status of Species," to assess the level of recovery of species and explore ways to improve the process. Besides providing a new tool to monitor the status of species, the Green Status moves away from what is often perceived as a "culture of despair" (one that highlights the negative aspects of the situation) and provides a more optimistic view that encourages conservation action at all levels.[6]

In addition to the creation of reserves and protected areas, many policies can effectively be implemented on a local or even a personal level. Land-use planning and zoning is one such tool; it takes into account population growth, regulates urban expansion, and protects agricultural areas, parklands, and other green spaces. Just as important is the concept of wildlife corridors, which preserve and establish links, such as tracts of woodland, between scattered habitats and allow animal populations to interact and find enough resources to survive. In India, for instance, there are close to one hundred elephant corridors, allowing unobstructed movements of herds.

In western Europe, wildlife networks existed in the past, when fields were separated by hedges and small forests, which provided an interconnected mesh along which wildlife could travel. However, the regrouping and consolidation of large agricultural areas that was widely implemented in the 1960s to improve mechanization and efficiency not only ruptured the wildlife network but also eliminated ponds and other precious water holes, as well as accelerating soil erosion through excessive runoff. By doing so, farmers unconsciously destroyed the natural support that insects, birds, and other small animals provided to their crops. Concerned countries have since realized what advantages were provided by the previous, hedge-rich arrangement, and the trend has now reversed to restoring "green belts" and patching the ecosystem back up. Even small garden owners can participate at their own level by preserving islands of wild grass and flowers amid or around their lawns to host and support insects, especially pollinators, such as bees and butterflies. Public authorities and large companies also play an important role by building wildlife corridors that bridge major throughways and railroads: not only overpasses for large animals, such as elk, but also underpasses for hare, frogs, and reptiles.

The Benefits of Biodiversity

In a world driven by profit, what might save biodiversity is what it can bring to the economy. Nations are mobilized against global climate change not because of some generous and altruistic awakening and commitment to protect future generations but for practical reasons: the cost of climatic disturbances, including agricultural losses and the price of natural disasters, already reaches 2 percent of the world's gross product—in the range of US$1 trillion per year—and could rise tenfold by the end of the century, according to the latest estimates.

Likewise, the reduction of biodiversity threatens to take a bite out of the world's economy. A report prepared for the Group of Seven (G7) countries meeting of environment ministers in 2019 estimated a worldwide loss, in US dollars, of anywhere between $4 trillion and $20 trillion per year in ecosystem services due to land-cover change (from agricultural to urban, for example) and losses of $6 trillion to $11 trillion due to land degradation. High-level global initiatives are being implemented as a result, such as the Bonn Challenge, aimed at bringing 350 million hectares (865 million acres) of degraded and deforested landscapes into restoration by 2030. Hope comes from the realization that benefits of restoration can far exceed the costs: in the Bonn Challenge, an estimated $7 to $30 of benefit could be generated for every dollar spent, not to mention that restoration work provides direct employment.[7]

Hence, besides philosophical and moral considerations, the preservation of species is about money, and there is nothing better than money to make governments and corporations react. One prime example of the role played by wildlife in the world's economy is our dependency on bees. Bees play a key role in the pollination of a variety of plants, boosting the productivity as well as the quality of major crops—cereal grains, fruits, and vegetables alike. The fact is that the populations of honeybees (*Apis mellifera*) and a number of other species of bees and bumblebees are collapsing worldwide. A study conducted in Europe shows that out of four hundred different bee species assessed over a third are experiencing a sharp decline in numbers.

The greatest threat experienced by bees is the destruction of their habitat through agricultural practices, such as abandoning hay production and no longer sustaining fields of wild grass and flowers, which are the favorite environment of bees. Individuals, one should remember, can help at their own level by leaving strips of grass unmown in their gardens or by mowing their lawn less frequently. Other negative factors are the excessive use of pesticides and herbicides, which kill bees directly or degrade their environment, and global climate change, which affects their habitat, through droughts or, inversely, too much rainfall.

The crash of bee populations has an enormous impact on the economy. The role of animals as a whole—bats and birds included—as pollinators of crops is estimated at over $150 billion per year worldwide: 10 percent of the total economic value of agricultural human food. Among pollinators, bees provide the largest share: in the United States alone, their contribution is estimated to exceed $15 billion annually. Hence, any decline in the bee population is felt immediately, as was the case in 2007, when farmers in the United States reported 70 percent fewer bee colonies and estimated the loss at $10 billion. Another bad year was 2014, when bee colonies crashed again, by 40 percent this time. One solution for farmers is to rent hives from beekeepers, but such migratory practices also spread viruses and other diseases across the bee population, as the hives are shipped around the country; this adds to the factors held responsible for population crashes (referred to as colony collapse disorder, or CCD). Pesticides, malnutrition, and loss of habitat are believed to be the other main factors.[8]

Remedies are actively being sought. In 2015, US president Barack Obama ordered a strategy plan to improve pollinator health, setting aside federal land for habitat protection—including planting wildflowers—and implementing measures to study and reduce wintertime bee losses. Environmental protection organizations also advocate reducing or banning pesticides, especially neonicotinoids.

Conversely, certain types of insects are harmful to crops and must be kept down to reasonable numbers: a service that is freely provided by insectivorous animals. Working in the shadows, bats offer direct support to agriculture: where they are healthy and active, less pesticides are needed, which not only benefits the environment but also cuts down expenses. Such bat-related savings are estimated to be worth up to $23 billion per year in North America alone. Besides their role in pest control, bats are also a major crop pollinator: over three hundred species of plants depend on the flying mammals for pollination, including mangoes, bananas, agaves, and the saguaro cactus.[9]

In North America, bats are endangered by a microscopic fungus, *Pseudogymnoascus destructans*, that proliferates on their nose while they hibernate: a disease known as white-nose syndrome. The skin infection is often lethal, threatening one bat species of extinction at a local level—the little brown bat (*Myotis lucifugus*)—and affecting many others, with population drops reaching 75 percent. The fungus thrives also in Europe, where it is carried by bats without causing mass mortality, and it is possible that it was introduced from Europe into North America, where bats had not developed the same level of defense against the parasite. Efforts are now being made to find treatments against the fungus, including fungus-fighting bacteria and even ultraviolet instrumentation at the entrance of caves to cleanse bats flying in and out of their refuge. Again, economic loss is a major incentive for setting up protection programs, since the bat population drop is estimated to cost half a billion dollars per year to the agricultural sector in the United States alone.

Better understanding the mechanisms at work in the ecosystem is another way to protect species, including those that are not currently at risk: it is wise to uphold the health of the ecosystem in a general sense, rather than to react when a species pops up on the Red List, often too late for countermeasures to be efficient. One engaging example is the support given to a near-threatened native falcon, the kārearea (*Falco novaeseelandiae*), in the vineyards of New Zealand. In 2005, falcons were introduced in a test area of the Marlborough wine region to control pest birds that eat or damage grapes, namely blackbirds, song thrushes, and starlings—all three introduced European birds—and the endemic silvereye. These pest birds were found to eat or damage by pecking 2.4 percent of sauvignon grapes and 3.4 percent of pinot noir. The introduction of the falcons, which caught or scared the birds away, resulted in damage from pecking dropping by 50 percent and damage from grapes being fully eaten dropping by 95 percent. Interestingly, pecking was less affected than full grape removal because pecking is mainly due to endemic silvereyes, which are less intimidated by kārearea falcons than European passerines are. The net result was yearly savings estimated, in US dollars, at more than $200 per hectare for the sauvignon blanc and more than $300 per hectare for the pinot noir. Extended to the entire Marlborough region, savings could reach up to $5 million a year, and over $50 million at the scale of the entire country. Moreover, falcons introduced in vineyards find a friendly and food-rich area to reproduce: a win-win situation.[10]

Few threatened or near-threatened species are as lucky as bees, bats, and falcons, attracting protection measures for the sake of the economy. Hopefully, governments, public authorities, private enterprises, and citizens will come to realize that at all levels of the ecosystem, ensuring the survival and well-being of species often brings about invisible and unquantifiable benefits that might only come to light when the species in question collapses. Who would guess that ants, often considered to be pests, boost wheat harvests by more than 35 percent in arid regions by digging galleries that improve soil permeability and water circulation? Ants also offer protection, for most crops, against harmful insects and fungi, and they synthesize a variety of precious molecules used in the pharmaceutical industry, notably to cure asthma and arthritis.[11]

Coral reefs constitute another realm of the ecosystem in need of protection. Because reefs support close to a quarter of all marine species, they are essential to the fishing industry—representing an estimated $6 billion of revenue a year—and therefore stand a good chance of being properly managed and preserved. If intelligently exploited, through the use of nondestructive fishing techniques, coral reefs generate each year 15 metric tons of seafood per square kilometer (0.4 sq. mi.); in Southeast Asia alone, reef harvesting generates over $2.5 billion per year.

Reefs also provide valuable services to the tourism industry—their attractiveness for scuba diving depends on their proper upkeep—and they protect coastlines by dampening the energy of waves and storms. Adding all these benefits together, coral reefs globally represent an economic worth somewhere in the range of $30 billion to $300 billion a year, so that economic potential alone should guarantee their proper management in the future, and the preservation of the many species they host. Efforts so far have focused on slowing down and eventually reversing the decline of coral reef area—which has shrunk 65 percent over the past half century—by restoring or transplanting coral beds and spreading their larvae, but nothing yet on the scale needed has been attempted, let alone been successful.[12]

Equally as important as coral reefs in terms of biodiversity, rainforests also call for close scrutiny and protection. They already benefit, in many cases, from national or international support in the form of parks and reserves, but the most effective way to protect them is to convince inhabitants, owners, and exploiters that properly managing rainforests and supporting their biodiversity is economically profitable. For one, rainforests hold thousands of species of medicinal plants that are of interest to the pharmaceutical industry, including more than two thousand that have cancer-fighting properties. One example is vincristine (a.k.a. leurocristine), extracted from the Madagascar periwinkle and used in chemotherapy.

More convincing yet are the short-term profits that can be derived from proper forest management. A study published in 1989, and still relevant today, demonstrates that clearing the Amazon rainforest to make room for cattle grazing creates land worth only $150 per hectare (2.5 acres) annually (plus up to $1,000 per hectare if the wood is sold—a nonrenewable, one-time profit), whereas managing the forest in a sustainable way, by harvesting its fruit and latex, can generate over $6,000 per hectare (1989 figures). Only through demonstrations of this sort will owners and exploiters turn away from destroying the rainforest and support instead its biodiversity and all it has to offer.[13]

Repairing the Environment

Besides limiting habitat destruction, population collapse, and the extinction of species, efforts can be directed toward reintroducing threatened species into areas from which they were removed, or into new habitats, in order to build a richer ecosystem: a strategy built on the understanding that animals and human beings are meant to coexist.

The reintroduction of wildlife focused first on predators: species threatened because their populations are usually low to begin with (being at the top of the food chain) and because they are hunted under the pretext that they attack livestock, wolves and wildcats being good examples. Such animals play an important role in regulating ecosystems by eliminating sick or genetically weak preys and keeping pest populations under control; as such, they are known are keystone species (see "Keystone Species" in chapter 6). Their reintroduction is frowned upon or opposed by herd keepers who suffer livestock losses as a result, so their role must be properly explained, and livestock losses accompanied by financial compensation.

Efforts to reintroduce threatened predators often meet with success, especially in Europe. Whereas countries like the United States and South Africa have built their wildlife strategy around national parks and reserves, separating animals from people, the tradition in Europe is to share the same space: a strategy adapted to the smaller areas typical of the continent, where predators need large roaming areas in order to survive and reproduce. What is particularly encouraging, as shown in one study, is that large predators are capable of roaming across relatively fragmented habitats, namely patchworks of forest and agricultural land where human presence is significant, twenty to forty people per square kilometer (0.4 sq. mi.).[14]

The success of human and wildlife cohabitation is partially due to protective legislation introduced and enacted by stable political regimes in Europe since World War II, and supported by public opinion and environmental organizations. Other helpful factors are the maintenance of healthy populations of deer, wild goats, antelopes, and smaller animals that act as preys, as well as the support given ranchers and shepherds, including the development of better paddocks and enclosures to shelter and protect their herds at night.

Besides predators, other species are being successfully reintroduced in the wild, although there is a bias toward the most popular and endearing animals: most reintroductions concern mammals (40 percent) and birds (30 percent). Reptiles, amphibians, fish, and invertebrates are lower down on the list, but their time will come, as it already has for turtles.

Large land tortoises have been found to play an important role in the ecosystem by dispersing the seeds of plants they ingest over long distances, despite their reputation as slow-moving creatures, because their intestinal transit time is also very long. Plans are being drawn to reintroduce tortoises in a number of Pacific islands in order to restore the diversity of the ecosystems. The Galápagos Islands already benefits from such a program, in view of the threatened status of their giant tortoises: as a result, the giant tortoise population has rebounded from a low of fifteen thousand in the 1970s to over twenty thousand in 2017 and counting. Despite their preservation, as such—many subspecies of Galápagos tortoises are still considered to be critically endangered, and some of these animals have been found to feed preferentially on invasive, rather than native plants, bringing added value to their reintroduction.

The reintroduction of species is a complex task, since one has to take into account not only their needs but also all the roles they play in their environment: the good health of an ecosystem is more important than the well-being of the individual species that make it up. Hence, there are situations when one should not focus on one endangered species in particular, at great expense of time and money, when it is possible to substitute another species that has a better chance of survival and can perform the same functions within the ecosystem equally well.

Another factor that must be taken into account is the spatial shift of temperature and humidity caused by global climate change, both latitude-wise and altitude-wise. The reintroduction of species should not be confined to their original habitats since these might soon become unsuitable. Many animal populations will need to be relocated outside their present occurrence areas to take into account global warming, and future land use should be planned accordingly.

Back from the Dead

One avant-garde strategy to fight extinctions is to resuscitate extinct species. The scenario explored by Michael Crichton in the novel *Jurassic Park* has indeed gone from science fiction to serious consideration. Biogenetic experiments are conducted today, although they often run into technical difficulties, such as the poor state of genetic material collected on the remains of extinct species. Improvements in collecting and storing DNA, however, are opening up a whole new list of resurrection candidates. Although dinosaurs are not yet in line, prospects are improving for recently extinct animals, like mammoths.

Woolly mammoths, as discussed in chapter 4, were part of the wave of megafaunal extinctions that took place at the end of the last glacial stage, some 15,000 years ago. Mammoths are special in that a number of corpses were frozen in Siberia's permafrost and are resurfacing today as the ice melts: one of the unexpected gifts of global warming. The experiment under consideration today is to collect DNA material from the nucleus of a well-preserved mammoth cell and splice it into the egg of a living female elephant in order to obtain hybrid offspring with mammoth-like traits, such as long hair, subcutaneous fat, and smaller ears adapted to cold climates.

Besides the issue of ethical considerations, mammoth DNA collected so far is much too deteriorated by thousands of years of deep freeze to offer any chance of success. Nonetheless, research is underway at Pennsylvania State University and at Harvard, and a biotechnology company, Colossal Biosciences, was founded in 2021 with the objective to revive the woolly mammoth and, more generally, to preserve extant endangered species through the use of gene-editing technology.

Another experiment in the making is Pleistocene Park, a nature reserve in northeastern Siberia, led by Arctic ecologists Sergey and Nikita Zimov. Its goal is to recreate an ecosystem similar to the one that prevailed during the last glacial stage by turning the moss and shrub-dominated tundra into a grassy steppe, the rationale being that such a steppe supported and was supported by the megafauna in a virtuous circle, and that reintroducing large grazing animals would operate the change. An expected side effect would be that vegetation change, along with the trampling and compacting of snow by large browsers, would help isolate the permafrost and slow down its melting, as well as its release of methane and other greenhouse gases into the atmosphere. The experiment, conducted on an area of 160 square kilometers (60 sq. mi.), began with the introduction of browsing animals such as horses, reindeer, and muskoxen in order to cut down shrubs and make room for grass; bison and camel followed in 2021.

The program has attracted media attention with the follow-up plan, if genetic manipulations are successful and the species brought back to life, to eventually introduce woolly mammoths, woolly rhinoceroses, and cave lions on site, giving full meaning to its name of Pleistocene Park. Other plans are being made to bring extinct species back to life, starting with the thylacine (a.k.a. the Tasmanian tiger), discussed in chapter 5, whose last representative died in 1936. Several thylacine pups are so well preserved in ethanol that, in 2017, a team of biologists was able to sequence the thylacine's entire genome: a blueprint that could be used to resurrect the species. One would still need to find a surrogate mother to provide a womb and pouch—another marsupial, like a wallaby or kangaroo, or an artificial womb and pouch. Next would come the difficult task of raising these fragile animals in captivity: remember, that's where the last thylacine died. Be that as it may, Colossal Biosciences announced in 2022 that, besides the woolly mammoth, it will also attempt to resurrect the thylacine.[15]

One final hurdle in reviving an extinct species would be to ensure that its members have enough genetic variability for the population to be viable; but if one masters the resurrection of a species from a genome, it should not be too difficult to program the required variability in individuals coming off the production line. Anywhere between fifty and five hundred distinct individuals would be required to provide a population diverse enough to make it viable.

Other emblematic and charismatic species are being considered in "de-extinction" programs. In New Zealand, high school students launched a project in 1999 to revive their school's mascot, the huia bird, which went extinct in the 1960s; although they received funding from an internet start-up, experts believe the DNA of stuffed huia birds kept in museums to be too deteriorated for the genome to be reconstructed.[16]

The same can be said for the giant moa bird, the remains of which were even less well preserved after its extinction over 500 years ago. One team, led by Japanese biologist Yasuyuki Shirota, did attempt to extract moa DNA for revival purposes, but without success. Interest in bringing the moa back to life follows the same logic as for mammoths in Siberia: before their eradication by human beings, the giant birds played a major role in shaping New Zealand's ecosystem through browsing and clipping the canopy, which kept the native bush healthy. Their resurrection would be welcome.

Reviving the dodo bird on Mauritius Island and the passenger pigeon in North America also rank high on the de-extinction list. In both cases, besides reconstructing viable DNA, the obstacles have to do with finding a parent species to act as a surrogate and obtaining a large enough population for the revived species reproduce efficiently—remember the high number of individuals needed for the passenger pigeon to thrive. Otherwise, all the invested time and money will simply lead to repeated extinctions of the animals in question.

Genetic engineering has met with some success with species and subspecies currently going extinct. The most encouraging results concerned a subspecies of wild goat, the Pyrenean ibex (*Capra pyrenaica pyrenaica*), that went extinct when Celia, its last representative, died in January 2000. Scientists had been able to collect some skin tissue from the female goat a year before her death and preserved the sample in nitrogen. Cloning was then performed by fusing cells of the tissue with eggs from goats that had their nuclei removed—in order to keep only the genetic contribution of the ibex—and implanting the fertilized egg into the womb of a domestic goat, who served as the surrogate mother. After many failures, one clone was born alive on July 30, 2003,

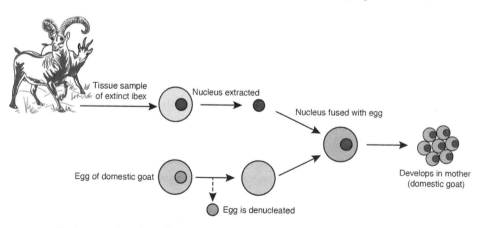

Strategy used to clone the Pyrenean ibex (*Capra pyrenaica pyrenaica*) in 2003. Somatic cells were taken from the last living Pyrenean ibex before its death, and the genetic material was fused with an oocyte from a domestic goat (*Capra hircus*). The egg was implanted into a surrogate goat mother for development. Adapted from 15ldavenport (Wikimedia), sketch of ibex by Joseph Wolf (1820–1899), licensed under CC BY-SA 4.0.

but died several minutes later of lung defects. Since that first set of experiments, no other attempt has been made to clone and resurrect the Pyrenean ibex.

Another partial success was achieved in 2013 with the cloning of Australia's extinct *Rheobatrachus* frog (discussed in chapter 5) using a related species to play the role of the surrogate mother. This time around, the cloned tadpoles survived several days, but again, results were not convincing enough to program further attempts to revive the extinct frog.

Experts do believe, however, that these results should encourage creating genetic banks of DNA, sperm, and eggs of the most threatened species: genetic material that can be used for artificial insemination, when animals are no longer able to reproduce naturally, or for cloning attempts in the future, when better technology becomes available.

The San Diego Zoo established the first cryogenic bank of this kind in 1972 and stores samples from close to one thousand species and subspecies. Half a dozen other institutions have followed suit, including the Audubon Nature Institute, paired with the University of New Orleans, and the University of Georgia's Regenerative Bioscience Center.

Establishing genetic banks and running cloning programs are very costly. A simple feasibility study to resurrect the Tasmanian tiger, undertaken by the Australian Museum, cost over $50 million before it was abandoned at the end of five years. The bottom line seems to be that our time, money, and energy are better spent upholding the quality and good health of our ecosystem and, in particular, curbing global warming, rather than focusing on desperate cases. Prevention is better than cure. It is everyone's responsibility to participate in the stewardship of our planet, to support those representatives and politicians who have a clear vision of what needs to be done, and to set an example at our own level, however modest, by adjusting our lifestyle and preserving our environment.

Is *Homo sapiens* an Endangered Species?

In our review of the biosphere, we assessed the status of a number of species but left one out in particular that we should be particularly concerned about: our own species, *Homo sapiens*. Like all other species, *Homo sapiens* emerged from a common bunch of cousin species to stand alone as a separate group of individuals, distinct enough so that reproduction became possible only within the group: the very definition of a species. The exact timing of this branching off from other members of the *Homo* genus, in the East African Rift, is a matter of ongoing discussion between experts: it is currently thought to have occurred around 300,000 years ago, from a *Homo erectus* ancestor.

Much more speculative is the fate of *Homo sapiens*. Our species can ultimately go extinct, putting an end to the entire genus, or weather all crises and remain virtually unchanged in its body plan for millions of years to come — or else evolve further, splitting into separate branches that acquire enough anatomical change to be considered new species.

Survival is a heavy responsibility for humankind because we are the only representative left of our group. Over the last 3 million years, our australopithecine forerunners and then the *Homo* genus delivered a number of different species, so that our evolutionary history should not be portrayed as a beeline inevitably leading to *Homo sapiens* as the result of some manifest destiny or intelligent design; rather, it should be seen as the expansion of a broad and bushy evolutionary tree, from which all branches broke off accidently, or through competition, until only one lone species remained.

Until about 50,000 years ago, three branches were still thriving: *Homo sapiens*, *Homo neanderthalensis*, and *Homo floresiensis*, the latter representing a very distant cousin that branched off the common *Homo* trunk around 2 million years ago. *Homo floresiensis* was one of the first members of our genus to leave Africa, migrating as far as Indonesia, where its 50,000-year-old skeletal remains were discovered in the island of Flores in 2003. Its extinction is broadly synchronous with the arrival of modern humans in the region. The two other finalists, *Homo sapiens* and *Homo neanderthalensis*, shared the stage in the Western Hemisphere, until the latter disappeared for reasons that are still unclear (see "*Sapiens* versus Neanderthal"), leaving us alone at the helm.

If we disappear as well, we will take down the entire evolutionary tree of the *Homo* genus with us, along with its cultural heritage and capacity to figure out the meaning of the universe. There are no substitutes in line after us to take over. This said, many animals, such as chimpanzees and dolphins, are endowed with elaborate forms of intelligence, so that there is still hope that one of them might develop a cerebral capacity allowing it to replace us, taking over our quest to understand the laws of physics and explore the origin and fate of the universe. At this point in time, however, we are the only species on Earth capable of probing and understanding the cosmos.

Other forms of intelligence might exist elsewhere in our galaxy, or in other galaxies, but there is no indication so far that this is the case. Hence, our survival might be particularly important in the grand scheme of things, if we are the sole conscious occupants of the universe.

In order to address the fate of our species, let us explore several possible scenarios. To begin, we must ask the right question and distinguish between the extinction of the human species and the collapse of our civilization, which does not necessarily lead to extinction. With respect to civilizations, history teaches us that nothing lasts forever: they eventually collapse under the weight of their own contradictions or are destroyed by some exterior agent. At present, the most obvious threats to our civilization are overpopulation, overexploitation of resources, and global climate change. But even faced with such perils, if our consumerist society was to collapse, there is a good chance that it would be replaced by another social order, without our species being overly affected.

Homo sapiens might indeed be especially difficult to eradicate. Our mammal ancestors did survive the end-Cretaceous mass extinction, arguably on account of their omnivorous, opportunistic feeding habits and their capacity to hide in shelters. The evolution of our *Homo* genus in East Africa, including the development of bipedalism and tool making, might even reflect its propensity to react positively to climate change, as speculated in chapter 4. On the other hand, if we focus on the more recent history of *Homo sapiens*, there is some indication that our species came close to extinction at least once in the past.

The Toba Eruption

The extinction of a species can occur as the result of a population crash—when the species reaches a critical population threshold below which any new source of stress might finish off the survivors or, in more general terms, the reproduction rate can no longer offset the death rate. It appears that *Homo sapiens* came dangerously close to such a critical bottleneck, based on the analysis of ancient and contemporary DNA—a method that makes it possible to estimate the demographic history of our species. In particular, our very low genetic diversity, compared to that of other animal groups like chimpanzees, indicates that the human population dwindled to a very low level, sometime between 100,000 and 50,000 years ago, before pulling out of its nosedive and climbing back to the high numbers witnessed today. Genetic calculations suggest that the human population, from which we are all derived, dipped below ten thousand individuals at the time of the bottleneck: about half the capacity of Madison Square Gar-

Sapiens *versus Neanderthal*

Our *Homo* genus has experienced one thought-provoking extinction: the disappearance of our close cousin, *Homo neanderthalensis*. The line of Neanderthals split from our own *Homo sapiens* branch some 800,000 to 500,000 years ago and developed in Eurasia: fossil skeletons show a stocky build with a barrel-shaped rib cage and other attributes well adapted to cold weather, proportionally shorter limbs than *Homo sapiens*, a sloping forehead, projecting nose, and a pronounced brow ridge. Neanderthals lived in small groups, were likely ambush hunters, made stone tools, engaged in primitive cave art, and collected crystals and seashells.

Homo neanderthalensis went extinct around 40,000 years ago, according to radiocarbon dates. Since modern humans (*Homo sapiens*) reached Europe between 45,000 and 43,000 years ago, there is an overlap of 3,000 to 5,000 years between the two species, and this has sparked debate about whether we are responsible for the extinction of our cousin.

Few other causes have been suggested. Bouts of abrupt climate cooling were once blamed, in particular a cold spell around 50,000 years ago that appears to have downsized human populations (see main text), but the morphology of *Homo neanderthalensis* seems especially well adapted to cold weather, making it unlikely that Neanderthals would succumb to a cold spell while modern humans sailed through it unscathed.

A more convincing theory has to do with genetic exchange. Much can be learned from the analysis and comparison of DNA from both Neanderthal remains and modern humans. The genome of European modern humans contains between 1 percent and 4 percent of DNA also reported in the Neanderthal genome, establishing that interbreeding took place between the two species and raising the possibility that Neanderthals vanished because their genes were absorbed

den in Manhattan, or a tenth of the capacity of Wembley Stadium in London.

Independent of the DNA study, geologists are aware of a catastrophic volcanic explosion that took place 74,000 years ago in Sumatra, Indonesia: by far the largest eruption that has occurred over the past 300,000 years, it released an estimated 3,000 cubic kilometers (700 cu. mi.) of magma, which is over ten times the output of the largest historical eruptions on record (Tambora in 1815 CE and Santorini around 1600 BCE). About a third of the magma was lofted into the atmosphere as fine ash, a veil that blocked sunlight and might have triggered a volcanic winter and overrun by those of *Homo sapiens*. According to DNA analysis, however, inter-breeding occurred mostly between 60,000 and 50,000 years ago in the Middle East; there was much less during the final overlap in Europe, when the Neander-thals truly disappeared, so that interbreeding is unlikely to be the main cause of their extinction. On the contrary, it is *inbreeding* that might have contributed to their demise. Because Neanderthals are thought to have lived in smaller groups than *Homo sapiens*, isolation during the cold spells that hit Europe might have led to excessive inbreeding and degeneration.

The most likely cause of the Neanderthal extinction, however, appears to be direct competition with *Homo sapiens* after the latter reached Europe and both groups began sharing the same space and scarce resources. As anthropo-logic studies have often shown, such competition often results in warfare. *Homo sapiens* was apparently far ahead in terms of armament—especially spears and other throwing weapons—and probably in terms of communication and war tac-tics as well.

Open warfare might not even have been necessary. By overhunting big game, *Homo sapiens* might have exterminated the main food source of Neanderthals. While *Homo sapiens* probably had no trouble using their superior technology to hunt smaller game, our cousins might have been unable to adapt to the change. One should also note that eliminating a competitor's food source is often con-sciously used in warfare, be it the scorched-earth strategy or the bison overkill performed by the US Army in the 1800s, done partially to cut off the food supply of Native Americans and force them off coveted land.

Most likely, not one single factor led to the extinction of the Neanderthal spe-cies but rather a combination of causes: the jury is still out.

over the entire globe. Initial calculations indeed suggested a 3°C to 5°C (5.4°F to 9°F) temperature drop caused by the eruption that would have lasted at least half a dozen years.

The idea of a connection between the Toba eruption, cooling, and the bottleneck of the human population was first suggested by science author and journalist Ann Gibbons in 1993; it was a concept supported and developed by geologist Michael Rampino, volcanologist Stephen Self, and anthropologist Stanley Ambrose.[1]

The Toba catastrophe theory, as it came to be known, was first challenged by further analysis of the DNA bottleneck itself. The decline of the human population is now thought to have begun closer to 150,000 years ago and to have proceeded gradually with no obvious acceleration at the time of the Toba eruption. In fact, archaeological evidence shows no sign of population reduction among human populations close to ground zero in Sumatra, or among our Neanderthal cousins in Europe for that matter.

The final nail in the coffin of the Toba theory came from the analysis of lacustrine sediment collected on the bottom of Lake Malawi, the southernmost lake of the East African Rift. The mud cores provide a precise record of the climate pattern in Africa at the time, and the year of the eruption is clearly marked by a layer of volcanic crystals derived from the Toba ash plume. Paleotemperature reconstructions based on the sediment suggest only a modest cooling of 1.5°C (2.7°F) following the eruption, and no sign of a lake overturn and redistribution of plankton and other small aquatic organisms, which would have occurred had the temperature fallen much below that figure. Microscopic fossils of plant cells in the mud also show little change in vegetation cover around the lake, the lower elevation grassy woodlands sailing through unscathed, while mountain forests were only slightly affected by the modest temperature drop.[2]

The fact that the Toba eruption had little to no effect on human demographics should not hide the fact that the *Homo sapiens* population did undergo a progressive decline, from over fifteen thousand individuals around 150,000 years ago to as few as five thousand individuals 30,000 years ago, before it slowly recovered.[3] It is likely our human ancestry underwent other population bottlenecks earlier in time, and some of our cousin branches might have gone extinct because of such crises. An interesting aside, however, is that bottlenecks can isolate gene pools that will favor and select particular traits, launching evolution into novel directions.

Whatever population crashes it experienced in the past, one would expect the human species to be out of danger today, in view of its overly large population. Its size has jumped from tens of thousands of individuals in prehistoric times to three or four million at the dawn of agriculture (10,000 years ago), around two hundred million in Roman times at the beginning of the Common

Era, one billion at the start of the Industrial Revolution in the late 1700s, and over eight billion at the time of this writing in 2023. The excessive growth of our population, however, comes with a number of pitfalls.

Pandemics and Bacteriological Warfare

If one abides by the IUCN criteria used to define threatened species, our population growth and habitat expansion should classify the human species in the "least concern" category. The situation is not as comfortable as it may seem, however, because our large population does not necessarily protect us from a global sort of danger. The first example that comes to mind is a pandemic, occurring naturally or set off accidently or intentionally by our own species.

Bacteria and other microbes have been invoked in the past to explain the extinction of a number of animal species, including dinosaurs.[4] The theory is still relevant today: we have seen how a number of amphibians in Central America have gone extinct under the attack of pathogenic fungi, and how the saiga antelope lost half its population in the spring of 2015 because of the opportunistic spread of a bacteria in the animal's bloodstream during an extreme climate event.

Human beings can also be decimated by a germ, as shown by the many pandemics that have struck our species in recent history. In the Middle Ages, a bubonic plague pandemic, known as the Black Death (1346–1352), took the lives of twenty-five million people in Europe—nearly half its population—and also affected Asia and Africa, with a worldwide death toll on the order of one hundred million. In terms of casualties, the twentieth-century Spanish flu tied that gruesome record with approximately one hundred million victims in two years (1918–1919), but since the world population was much larger at the time (close to two billion), that figure represents at most 5 percent of the total population. The toll would have been much higher, however, had the virus been more lethal, since half the world's population was probably infected.

The Spanish flu demonstrated that, in the modern world, rapid and large-scale population movements can blow up an epidemic, which otherwise might have only affected a restricted region, into a global-scale pandemic capable of threatening the entire species. In the earlier example of the Black Death, the epidemic started in Central or East Asia and spread to Europe and Africa by way of merchant ships, which carried rats and their fleas across the Mediterranean. As for the Spanish flu, despite its name,[5] it initially developed in the United States (the first recorded cases were in Kansas), spread from military training camps to US Army camps in Europe during World War I, and propagated to fighting troops of all nations involved.

Today, population movements have reached an unprecedented scale, so that the propagation of a pandemic can be of frightening efficiency, to the point of threatening our civilization and perhaps even our species. In order for this to happen, three conditions must be met: the bacteria or virus must be extremely contagious, it must be deadly, and it must be difficult or impossible to eradicate or neutralize. Our species has been lucky so far, considering the Spanish flu fulfilled only two conditions out of three—it was extremely contagious and impossible to stop at the time—but it was deadly in only 10 percent of the cases.

More recently, the COVID-19 pandemic, which began in December 2019, also fell short of meeting all three conditions. It did meet the first: it was extremely contagious. Although six hundred million cases were reported over a three-year period (2019–2022), which is less than 10 percent of the world population, many more people were infected and their cases not reported, or they showed no symptoms: an estimated three and a half billion people, which is closer to 50 percent of the world population. The mortality rate for those who contracted the disease, however, was relatively low: seven million casualties by the end of 2022, although many deaths have gone unreported or attributed to other causes, so that some experts believe the real number is closer to three times the published figure (eighteen million casualties). Even using this projected number, the casualty rate boils down to 0.2 percent of the world population, far from the 5 percent reached during the Spanish flu and the 30 to 50 percent figure reached during the fourteenth-century Black Death. Finally, the COVID-19 pandemic, contrary to the two previous diseases, was considerably mitigated by hygiene recommendations (face masks and hand washing), social distancing, confinement and quarantining, and the speedy development of vaccines.

It is interesting to note the differences from country to country, with respect to infection and casualty figures, due to a number of factors such as hygiene, background health status of the population, rapidity of response to the crisis, and different strategies adopted by governments. New Zealand scored the lowest COVID-19-related casualty total (0.001 percent of the population; i.e., one casualty per one hundred thousand inhabitants) due to geographical isolation, early lockdown, and travel restrictions, whereas the highest casualty figures (in Peru and Bolivia) are closer to 1 percent of the population—one thousand times higher. The United States and the United Kingdom rank in the middle range, with around 0.3 percent of the population killed by the disease.

The COVID-19 pandemic served as a dress rehearsal for future pandemics in terms of what strategies to deploy, how fast we should deploy them, and how efficient they are. Nature is turning out new viruses at a speedy rate, in particular because of increasing contact between human beings and birds, bats, pigs, and bush animals, including monkeys. Birds in particular are carriers of a great number of viruses and can transmit them to poultry, which in turn pass them on to livestock mammals, such as pigs, which serve as incubators and develop new versions of the initial viruses; because of the closeness of the porcine genome with our own, the disease can then be easily passed on to human beings, hence the many cases of avian influenza and swine flu. In the case of the COVID-19 epidemic, for example, its outbreak is convincingly traced to a Huanan market in China where vendors sold live mammals, including red foxes, hog badgers, and raccoon dogs, one or several of which were most probably the carriers of the disease (working backward, the wild mammals were probably contaminated by bats).[6]

As was the case for COVID-19, many such viruses have a low lethality, so far. The 2009–2010 epidemic of the H1N1/09 influenza, for example, known as the swine flu and originating from the same virus family as the Spanish flu, caused "only" three hundred thousand casualties worldwide (approximately 0.005 percent of the population); the 1968 "Hong Kong flu" (H3N2) is estimated to have decimated closer to 0.1 percent of the world population, in the same range as the COVID-19 pandemic.

There might come a time, however, when a pandemic will break out that not only is very infectious—spreading very fast within the population, as has been the case in the past—but also much more lethal than previous occurrences, and much more resistant to possible vaccines and medication. Medical research attempts to stay one step ahead of viral mutations and developments, which leads high-security laboratories to cultivate viral strands and to experiment and analyze possible mutations.

The paradox is that such research can lead to accidents leaking dangerous viruses into the environment. There are about sixty laboratories across the world, classified biosafety level 4 (BSL-4), that handle viruses and bacteria that could cause fatal diseases if released in the environment; they have tightly controlled air flow and decontamination airlocks, and their access is restricted. They have experienced several accidents, however, including the deaths of two scientists at the Vector Institute in Koltsovo, Russia, one in 1998 and one in 2004, who became exposed to the Marburg and Ebola viruses, respectively. More recently, China's Wuhan Institute of Virology had long been suspected of playing a role in the spread of the COVID-19 virus, since the laboratory specializes in the study of coronaviruses, including their transmission from bats to people; it was actually the first laboratory to publish the genome of the COVID-19 virus in February 2020, three months after the disease broke out. The theory that the virus leaked out of the lab, accidently or intentionally, is judged by experts and investigators to be extremely unlikely, however, and a natural origin of the virus, transmitted from animals to humans, is by far the favored explanation, as previously discussed.[7]

Besides the eventuality of a natural or an artificially sparked pandemic, there is the added threat of lethal viruses being stolen by terrorists or used by rogue states to conduct bacteriological warfare. The concept of bacteriological warfare is not new. One of the first occurrence on record was during the fourteenth-century Black Death plague, when Mongol forces attacking a Genoese trading center in the Crimean Peninsula reportedly catapulted plague-infested bodies over the walls of the city. Another infamous example, in 1763, is the distribution by British troops of smallpox-infected blankets to Native Americans, although the strategy failed because smallpox doesn't spread easily via objects (however, it did spread from person to person, greatly affecting the Native American population).[8] One century later, in 1863, Confederate agents similarly planned to send yellow fever and smallpox-infested clothing to Union troops, and to President Abraham Lincoln himself, but the plot also failed.

Despite international meetings and treaties that repeatedly outlawed bacteriological warfare (most notably during the 1899 Peace Convention at The Hague), biological weapons were tested during World War I and used by Japan against China during World War II, by poisoning water wells with cholera and typhus germs and dropping plague-infested fleas from planes over Chinese cities. Today, however, states have abandoned research programs of bacteriological warfare, considered the "poor man's" weapon relative to high-tech alternatives like the atomic bomb, although the threat of terrorists or sects using smallpox or anthrax to conduct an attack on a local scale is still a concern, and germ handling and experimentation are closely monitored.

All in all, it appears very unlikely that the human species is at risk, with respect to viral or bacteriological infection. Civilization rather than humankind, however, is more fragile in this respect, as demonstrated by the economic chaos triggered by the COVID-19 pandemic. In the future, an equally fast-spreading but more lethal infectious disease would destabilize our civilization on a much larger scale.

Nuclear Warfare

Even if a catastrophic pandemic causes our civilization to collapse temporarily, major agricultural and industrial infrastruc-

Nuclear weapon test (Operation Crossroads), Bikini Atoll, July 25, 1946. Today's nuclear warheads are twenty times more powerful. Photo by US Department of Defense.

tures would be spared, so that survivors would have the means and the tools to continue business as usual after recovering from the crisis. The situation would be very different, however, in the aftermath of a global nuclear war.

The prospect of a nuclear war has been looming since the end of World War II and the arms race between superpowers. Despite treaties to curb the number and the power of atomic bombs, an estimated 9,400 nuclear warheads were operational[9] in military stockpiles as of 2022, 85 percent of which are shared equally between the United States and Russia (about 4,000 each), with China a distant third (350 bombs), followed by France, the United Kingdom, Pakistan, India, Israel, and North Korea. Although the number of nuclear warheads has dropped considerably since an all-time high of 65,000 in 1985, the combined world arsenal still represents today over 10,000 megatons of TNT. By means of comparison, the total power of all bombs used during World War II (Hiroshima and Nagasaki included) adds up to about 3 megatons of TNT.

The threat of nuclear war ebbs and flows, with very tense close calls, during the Cold War—the 1962 Cuban Missile Crisis and the Euromissile crisis (1977–1987)—and during Russia's invasion of Ukraine in 2022. Considering the standoff between superpowers (deterrence) is based on the capacity of immediate response to a threat, missiles are on hair-trigger alert, leading to a situation where logic and restraint might one day not apply, and the onset of a nuclear war could occur simply by mistake, be it human or technical. In moments of tense political clashes, such as the one that put Russia at odds with the Western world in 2022 over the invasion of Ukraine, some experts estimate the probability of a nuclear exchange around 10 percent over the course of a year. At the time of the 1962 Cuban Missile Crisis, President John F. Kennedy reportedly estimated the chances stood between 33 and 50 percent. Even in times of worldwide political stability, most experts judge the likelihood of a nuclear war breaking out somewhere on the planet, on any given year, to be more than 1 percent.[10] Carrying the statistical analysis to its logical conclusion, that would mean that nuclear warfare is virtually inevitable over the course of a century. The only way to back out of this dead end, so to speak, would be to bring the stockpile of nuclear weapons down to zero.[11]

To illustrate the many possible actions and misinterpretations that could lead to the accidental triggering of a nuclear war, consider the following example. On September 26, 1983, during the Euromissile crisis of the Cold War, which had reached a climax after the accidental downing of a passenger plane carrying Americans (Korean Air Lines Flight 007), a Soviet satellite early warning system mistakenly reported the launch of half a dozen US missiles toward the Soviet Union. The Soviet officer on duty, Stanislav Petrov, made the cool-headed decision to consider the satellite data to be a false alarm, and he convinced his superiors not to pass the information higher up until ground radar confirmed the attack was real. Had the early, faulty information been passed on, it might very well have prompted Soviet leader Yuri Andropov to launch a full-out nuclear retaliation toward the United States, triggering World War III.

On January 25, 1995, there was another incident, this time escalated up the chain of command up to Russian president Boris Yeltsin. In this case, the detection of a rocket launch was real: it was a Norwegian sounding rocket aimed at studying the northern lights, but when detected by a Russian radar, its trajectory resembled a trajectory that might be used by American missiles to attack Moscow. President Boris Yeltsin actually activated the key of his "nuclear briefcase" and stood ready to launch a full-out nuclear attack against the United States. The tense moment ended when the sounding rocket fell to the ground far from the Soviet Union.

In light of these close calls—and the several others that have occurred—a nuclear war scenario is far from absurd, and it is worth contemplating. An all-out nuclear war would probably lead to the collapse of human civilization: besides killing a sizable portion of the population and plunging civilization in chaos, a nuclear exchange would cause a major cooling of the world's climate and the collapse of agriculture: a scenario known as a "nuclear winter."[12]

Even a limited, regional conflict would set off a climate upheaval, as outlined in a simulation run by Jonas Jägermeyr and his collaborators, published in 2020. The scientists ran computer simulations reproducing a limited nuclear war, in this case between India and Pakistan (about one hundred Hiroshima-class atomic bombs detonated). Besides decimating the populations involved (tens of millions of victims), the blasts would ignite enough fires to loft more than 5 teragrams (5 million metric tons) of black carbon soot into the atmosphere. The end result would be an average temperature drop across the entire planet on the order of 2°C (3.6°F)—the coldest surface temperature in the last 1,000 years—that would last for at least five years, compounded by a luminosity drop of up to 5 percent and a precipitation drop of 8 percent. Essential crops, such as maize, wheat, and rice, would experience a caloric production drop of roughly 10 percent worldwide (more than twice the largest anomaly on record since global monitoring of food sources began in 1961); in some regions, like the United States, the production drop might reach 20 to 40 percent, mimicking the dustbowl situation of the 1930s. According to the simulation, it would take 10 to 15 years for the crops to recover.[13]

The repercussions of such a collapse would include a global breakdown of trade, inflated food prices, social unrest, and increased hunger and health issues around the world. But in this regional conflict scenario, civilization would be unlikely to collapse, and certainly the human species would not.

The situation would be quite different in the case of a nuclear war between superpowers like the United States and Russia, which could involve several thousand bombs. According to one study, such a full-out nuclear exchange would directly kill—by blast, heat, and fire—somewhere around five hundred million to one billion people. The effect of global radioactive fallout, on the other hand, is often exaggerated and would only raise the casualty figures by a few percent. The real damage would again come from a nuclear winter, but on a much larger scale than in the regional war scenario. Carbon soot lofted into the atmosphere would be thirty times more abundant—on the order of 150 teragrams, rather than the 5 teragrams in the previous example—and would trigger a temperature drop of approximately 10°C (18°F): twice the temperature drop necessary to flip from an interglacial stage, like today's, to a runaway glacial stage. Soot would drop out of the atmosphere on a timescale of 10 years, but severe damage would have already occurred. Agriculture would crash on a major scale. Simulations predict that summer temperatures would dip below freezing in the heart of continents like North America and Eurasia, and crops would be decimated at midlatitudes, while the lack of rainfall in the tropics, combined with cold temperatures, would be equally devastating. Global food production would be cut by more than 90 percent (up to 98.9 percent in the United States, according to the simulations), causing worldwide famine and several billion extra casualties. But even then, probably only a quarter of the world's population might survive.[14]

There is no doubt that, at this stage, our civilization would totally collapse. The biosphere itself would suffer a major blow, with the drop in temperature and sunlight killing off a great proportion of plants on land and plankton at sea, triggering a breakdown of the entire food chain. The good news is that extinction of the human species would still be unlikely, even in an all-out nuclear exchange using up the entire arsenal of atomic bombs. Physicist and disarmament activist Joseph Rotblat (1908–2005) estimated that for this to occur, at least ten to one hundred times more power is needed—and even that might be far from sufficient, if we recall that the end-Cretaceous asteroid impact delivered ten thousand times more power than our entire nuclear arsenal, and yet our mammal ancestry and 25 percent of all species on Earth survived. It might take more than nuclear warfare to do us in, but human beings are never short of imaginative ideas to set up their self-destruction.

Artificial Intelligence and Extermination

While the human species appears incapable of ensuring its own extinction with "ordinary" atomic weaponry, it has opened up new possibilities by developing nanotechnology and artificial intelligence.

Nanotechnology consists of building tiny functional devices capable of self-replication and dissemination. The question is, what would happen if such devices were programmed by ill-intentioned people, or simply escaped our control, and multiplied to darken our skies, devastate our crops, or devour all organic matter, human beings included? Nicknamed "gray goo" in science fiction, the concept relies on exponential growth and need not be malicious to wreak havoc in our civilization. One would expect countermeasures to be fairly straightforward to devise and implement in order to quell the threat; the main problem here is speed and efficiency of reaction, because exponential growth is ruthless, as we know from the spread of viruses.

Things get worse when artificial intelligence comes into play. Since the invention of computers, the number of transistors in a processing chip approximately doubles every two years. From 1956 to 2015, this exponential growth has resulted in a one-trillion-fold increase in processing power. The computer data crunching and output of solutions, suggestions, and strategies at lightning speed surpass anything that human beings can generate on their own, as illustrated by the fact that supercomputers have become unbeatable at the game of chess since the first victory of IBM's Deep Blue computer against world champion Garry Kasparov in 1997; two decades later, even mobile phone programs are powerful enough to run at the grandmaster level. As for the more complex and creative game of Go, it too has been assimilated by computers, and its human champions systematically beaten since 2017, causing South Korean grandmaster Lee Se-Dol to retire two years later, stating, "I'm not at the top even if I become the number one: there is an entity that cannot be defeated."[15]

Board games are one thing, but artificial intelligence can be programmed just as well to devise war games and unbeatable weapons. At the 2021 Convergence Conference, organized by the Swiss Federal Institute for NBC (nuclear, biological, and chemical) Protection (the Spiez Laboratory), the US biotech company Collaborations Pharmaceuticals demonstrated how drug-design programs assisted by artificial intelligence and intended to benefit human health could be used to harm humanity instead. Feeding chemical toxicity data into their drug-design algorithm, company scientist Fabio Urbina and his team obtained at the end of a six-hour run a list of forty thousand lethal compounds, including new ones predicted to be more toxic than chemical warfare agents currently known. In the paper describing their experiment, the authors not only showed that "by inverting the use of our machine learning models, we had transformed our innocuous generative model from a helpful tool of medicine to a generator of likely deadly molecules," but also raised the issue that "importantly, we had a human in the loop with a firm moral and ethical 'don't-go-there' voice to intervene; but what if the human were removed or replaced with a bad actor?"[16]

The question is not only ethical: it could become a matter of survival for humanity. As long as artificial intelligence is controlled by its creators and programmers with good intention, it can indeed help solve problems faced by civilization. But what if machines—with the help of bad actors or, more provocatively, entirely on their own through some internally generated logic—conclude that it would be judicious to eliminate human beings because we compete for their resources and limit their efficiency and growth potential or, on a different scale, threaten the future of the biosphere?

The question was addressed by science fiction Arthur C. Clarke and filmmaker Stanley Kubrick in their 1968 epic, *2001: A Space Odyssey*, in which the supercomputer of a crewed spacecraft bound for Jupiter decides to kill the crew, judging it inadequate to carry on the mission successfully. Likewise, an artificial intelligence could logically decide to eliminate the human species, if it is programmed to solve the problems of the planet and computes that humankind is the most damaging element of the system.

Considering the fact that great world powers, including the United States, Russia, and China, have developed killer robots to replace live soldiers in the battlefield (drones and other automated vehicles) and that their instructions of what enemies to eliminate simply consist of selecting a set of parameters, an artificial-intelligence program could very well conduct warfare against us and use also nanotechnology in the process. According to the Future of Humanity Institute (Oxford University) and the Global Challenges Foundation's report *12 Risks That Threaten Human Civilisation*,[17] a destructive takeover by artificial intelligence is probably the most problematic threat faced by human civilization, considering, as we saw with the game of chess, once artificial intelligence masters the rules of the game, it will be impossible to beat—and it will be wise enough, once it starts considering humankind a nuisance, to hide its calculations and intent until it is sure that, when it implements it plan, it will win by knock out.

The Threat of Super Volcanoes

In its report, *12 Risks That Threaten Human Civilisation*, the Global Challenges Foundation lists extreme climate change, ecological catastrophe, a global pandemic, nuclear war, nanotechnology, and artificial intelligence as the most obvious threats to humankind. It also lists two exogenic risks, coming from outside our biological and human system: a super-volcano eruption and a major asteroid impact.

The threat of super-volcano eruptions—eruptions unleashing more than 1,000 cubic kilometers (240 cu. mi.) of magma—has already been discussed in the section on the Toba eruption, 74,000 years ago, for which there is no convincing evidence of an associated climate breakdown, or of a noticeable dip in the human population at the time.

Based on the geological record, experts estimate that large eruptions of that kind occur once or twice per million years, so that the *Homo* genus has weathered at least half a dozen of these without going extinct. Although they appear innocuous at the species level, such eruptions would certainly affect agriculture and civilization, so that we should be prepared to deal with the next one. When could it happen? Much has been written about the present threat of the Yellowstone and Long Valley super volcanoes, in Wyoming and California, respectively, but the fact is that with all volcanoes put together, the probability of a super eruption doesn't rise much above 0.01 percent per century across the world. This doesn't mean that such an eruption won't happen next month—probabilities only reflect long-term trends—but compared to nuclear war, the odds are low.

Equally improbable, but more lethal if it does happen, is the onset of a "great igneous event," of the kind that spreads *hundreds of thousands* of cubic kilometers of lava and ash over an interval of hundreds of thousands of years, the last event being the Columbia River Basalts, around 16 million years ago. The earlier Deccan Traps of India (66 million years ago) are still believed by some to have triggered the end-Cretaceous mass extinction, or at least to have contributed to it, but no convincing evidence supports that hypothesis (see chapter 2). More puzzling are the Siberian Traps that erupted 252 million years ago, and their coincidence with the end-Permian mass extinction, upholding the theory that the crash of the biosphere was caused by an outpouring of eruption-related carbon dioxide and methane (see chapter 1). Scientists will undoubtedly learn a great deal about climate change and lethal feedback loops by unraveling the mysteries of the end-Permian mass extinction and the Siberian Traps, but the probability that a super eruption of that kind will threaten human civilization is low, considering the long time intervals between events: 25 million to 50 million years on average, so that the odds of a trap-like eruption breaking out within the next millennium are no higher than 0.004 percent.

The Threat of Asteroids

The odds of an asteroid impact bringing down the biosphere and extinguishing humanity should also be assessed. Such an impact-triggered collapse of the biosphere happened at least once before, with the demise of the dinosaurs and the extinction of 75 percent of all species 66 million years ago, marking the end of the Cretaceous period. The irony is that the catastrophe opened the way for the ascent of mammals, primates, and ultimately the human species, but our luck could turn around: a future blow could wipe out humanity in turn and cause a new reshuffling of the ecosystem.

The frequency and magnitude of impacts are relatively well-understood, through the study of impact craters on the Earth and the Moon and the tally in space of Earth-crossing asteroids and comets, known as near-Earth objects (NEOs). For example, an impact by an asteroid that is 500 meters (1,640 ft.) wide, blasting a crater close to 10 kilometers (6 mi.) in diameter and unleashing a power equivalent to an all-out nuclear war, should occur somewhere on Earth every 50,000 years or so. The odds of such an impact are thus much smaller than those of a nuclear war, which is five hundred times more probable, according to what we saw earlier. If we are able to defuse the threat of nuclear war (as well as that of artificial intelligence), an impact of that size would then become the next-in-line most significant threat to our civilization, with a 0.2 percent probability of occurring within any given century, which is twenty times the odds of a super-volcano eruption occurring in the same time interval (0.01 percent per century).

The odds of a larger impact are smaller yet, since large asteroids are much rarer in space—only one thousand near-Earth asteroids are larger than 1 kilometer (0.6 mi.) in size. It would probably take a 2-kilometer (1.2 mi.) object, blasting a crater at the Earth's surface some 30 to 40 kilometers (19 to 25 mi.) in diameter, to completely break down global climate, freeze crops, and starve humankind, leaving few survivors. Such an impact is estimated to occur every 1 million to 2 million years on average, which is pretty much the time scale of the human species. In other words, it might eventually occur down the line and become the event that terminates our species, but in the short term, it is not a pressing concern.

Finally, an impact of the same magnitude as the end-Cretaceous Chicxulub event, a 10-kilometer (6 mi.) asteroid or comet, which would have a high chance of completely exterminating *Homo sapiens* as well as the vast majority of species on Earth, is estimated to occur every 150 million years on average. Such low odds apparently place such an event completely outside our sphere of concern, since the probability of it happening in the next century is less than one chance in ten thousand, but it should always be remembered that however small the odds, such an event will occur at some point, and that its exact timing is totally random and outside our control. It could happen in 100 million years, or it could happen next year.

What we might be able to control is mitigation. Paradoxically, it might be easier to neutralize a threat from outer space than a threat on our own turf: there is no known way to prevent the eruption of a super volcano, but there are ways to prevent a cosmic impact from happening. One big advantage we have on our dinosaurian predecessors is our intelligence, coupled with high technology. Dinosaurs had no idea what was coming and no way to prevent it from happening, whereas the human species has reached the point where it understands the threat from outer space and has built up the capacity to fight it.

Interestingly, we have only recently reached that stage: the study of impacts and their consequences, the detection of NEOs and the development of means to destroy them, all took place over the past 75 years.

The tallying asteroids and comets and calculating their trajectories and chances of collision with Earth are now well advanced. With respect to the interception of an incoming object, the irony is that the very weapons designed by our species to destroy itself could be used to save it, by detonating nuclear bombs above an asteroid or comet and, through the shove of the detonation, deflecting its trajectory sufficiently so that it avoids our planet.

The irony comes with a twist, however. Experts believe that the only way to avoid a nuclear war is to dismantle all nuclear weapons. By doing so, humanity would avoid self-destruction, but it would lose the ability to intercept incoming space rocks with those tools.

Given the high odds of a nuclear war and the low odds of a major impact within humanity's lifetime, dismantling nuclear weapons is by far the best way to go if we wish to survive as a species. There will always be ways to devise "soft" means to deflect asteroids—space tugs, ion beams, or focused solar energy—without resorting to weaponry that could be used against us. Kinetic energy imparted by a large enough mass, through simple collision, is one solution explored and tested by way of NASA's Double Asteroid Redirection Test (DART) mission, which crashed a 570-kilogram (1,260 lb.) probe into small asteroid Dimorphos (160 meters, or 525 ft., in size), on September 26, 2022. The change of velocity imparted to the target (in this case a breaking of its orbital motion around larger asteroid Didymos) amounted to approximately 2 centimeters (0.8 in.) per second.[18]

Artist rendition of the asteroid deflection test performed by NASA's DART probe on asteroid Dimorphos, September 26, 2022. Illustration by NASA / Johns Hopkins APL / Steve Gibben.

If an asteroid of that size were to threaten Earth, such a collision might impart enough velocity to the object for it to drift past Earth by enough of a distance to avoid collision, but only after 10 years of drifting sideways at that rate of 2 centimeters (0.8 in.) per second. In other words, humanity would need a 10-year lead to launch a strike against such a threatening small object, and more lead time, or a higher impacting mass, if the asteroid was larger (and thus more difficult to steer). So, beyond these initial experiments, there is still much work to be done.

Mars as a Lifeboat

Were our planetary defense system to fail—against a pandemic, nuclear war, or an asteroid impact—there would still be one rampart left against extinction: implanting offshoots of the human species on several planets rather than one, following the wise saying of not putting all your eggs in one basket.

After originating in Africa's East African Rift, the human species spread over all continents, motivated by curiosity and necessity. Technology now makes it possible for humankind to settle even more distant lands, and in particular our neighboring planets. Most are uninhabitable—Mercury is airless and too hot, Venus a suffocating waterless greenhouse nightmare, Jupiter and the other giant gaseous planets landless and too cold—but the one exception is planet Mars.

Within six months travel from Earth, smaller than our world but with an area comparable to that of all our continents combined, the red planet has great potential in terms of habitability. Its surface temperature is comparable to that of Antarctica: $-60°C$ ($-76°F$) on average, but up to $30°C$ ($86°F$) at ground level during summer afternoons at the equator), with the possibility for settlers, at least in theory, to thicken its carbon dioxide atmosphere and bolster its greenhouse effect. Mars has vast reservoirs of water, mostly in the form of polar caps that represent a global layer of water 35 meters (115 ft.) deep if melted and spread over the entire planet, with perhaps more water locked up in the crust, and its abundant raw materials can be used for construction (clay, sand, plaster) and for a burgeoning industry (iron, carbon, silica).

However, Mars should not be considered an auxiliary planet capable of absorbing a large spillover of population from Earth, were our world to be threatened or destroyed by climate change, a nuclear war, or any other planetary breakdown. Limitations as to the size of a Martian population are mostly economic, because of the cost of space travel. Currently, with rockets of the Delta or Vulcan class, landing 1 kilogram (2.2 lb.) of payload on Mars costs approximately $250,000. Travel costs are bound to fall significantly in the future, however; based on developments at SpaceX, notably the Starship program, Elon Musk believes the cost of shipping to Mars could drop a thousandfold, and he is confident that in the not-so-far future, moving to Mars would cost less than $500,000 per person, freight included, "low enough that most people in advanced economies could sell their home on Earth and move to Mars if they want."[19] Whether this is an over-optimistic quote remains to be seen, but the bottom line is that only wealthy people will be able to move to Mars in the foreseeable future. The red planet is definitely not a lifeboat for humanity as a whole; however, it is entirely conceivable that it will be settled by an elite, be it a wealthy or a scientific elite or a combination of both. Elon Musk speculates that with a fleet of one thousand Starships, up to a million people could be moved to Mars in the course of the century (one hundred thousand people during each launch window, every two years or so),[20] but even then, this figure would still represent only 0.01 percent of the Earth's population.

Though it's short of moving large populations between planets, starting a small colony on Mars is technically feasible. Elon Musk predicts a first landing of his Starships on Mars around 2029, although schedule drifts are customary in space exploration. Time will tell.

One interesting aspect of a Martian colony, or of any colony for that matter, is how big it should be to start a new, genetically healthy branch of human civilization. Experts estimate that the founding population should consist of a total of at least 150 men and women for the colony to avoid the risk of consanguinity and hereditary disorders. Better yet, an initial population of 500 to 1,000 would cut down the chances of genetic drift—the reduction of genetic variation over time—and promote instead enough genetic mutations to bolster diversity and keep the population flexible and ready to face any adverse evolutionary pressure.[21]

On the other hand, a genetic drift reducing the diversity of a Martian colony is not necessarily a bad thing: it could lead to the enhancement of specific features and give birth to a novel branch of the human species. In the long run, Martian conditions could favor new traits, as might the lower gravity field (0.38 g; i.e., 38 percent of Earth's gravity) that would undoubtedly affect muscle mass, bone structure, and the cardiovascular system—changes that could be engraved into the human gene pool by favorable mutations and perpetuated down the line.

How long would it take for the population of a Martian colony to separate sufficiently from the terrestrial main branch in order to create a distinct subspecies or even species? No one can tell at this point, but regardless of the outcome, the eventual extinction of *Homo sapiens* on Earth would not necessarily mean the end of humanity as a whole.

Silence of the Stars

Let us take the space-faring speculation one step further. After colonizing Mars, *Homo astronauticus* could sail from star to star, exploring and exploiting new worlds, driven by curiosity and necessity. Some thinkers believe that such galactic wandering of an intelligent, technological species is probable, and even inevitable. Such speculation, however, leads to a troublesome conclusion: given the great number of planets in the galaxy—probably several hundred billion—and even if one assumes very low odds for the emergence of life and for the development of intelligence, our galaxy should, by now, have entirely been explored and colonized by civilizations older and more advanced than ours. However, to date, we have found no trace of them.

This enigma was first brought up by Italian American physicist Enrico Fermi during a casual discussion with colleagues in 1950, at the end of which he launched the now-famous question "But where is everybody?" followed by speculations as to the probabilities of planets, life, intelligence, and the rise and duration of a technological civilization. Known as the Fermi paradox, the absence of any sign of conscious existence outside of Earth can be explained if extraterrestrial intelligences are much rarer than our back-of-the-envelope calculations seem to suggest, or if they are extremely discreet or their lifetime is very short.

Juggling with all these factors is a thought exercise framed by what is known as the Drake equation, named after American radio astronomer Frank Drake, who formulated it in 1961. Designed to structure discussion by breaking down the problem into its constituent questions, the equation leads to an estimate of the number of advanced civilizations present in the galaxy at any given time. Using it backward can also enlighten us on our chances of survival as a species.

If we plug in ballpark numbers to quantify the unknowns of the equation, such as the probability of planets around other stars and their potential habitability in terms of temperature and water (figures that are now reasonably well known), as well as the probability of the emergence of life, intelligence, and technology (unknown, but many experts speculate it to run in the 0.01 to 1 percent range for each of the three figures), then the last factor left to solve the equation, and explain why we do not detect any other civilization in our galaxy today, happens to be the average lifetime of civilizations like our own.

When I play this game myself and plug in my own numbers, my hunch is that an advanced, communicative civilization apparently lasts less than 100,000 years, which I take as the upper bound for the lifetime of our own civilization, from the time it reaches its technological and self-destructive status. This is one way to explain why we do not detect anyone else out there in the galaxy. Because of the self-destructive nature of advanced civilizations, there is only one that is active and detectable at any given time, and currently, we are that one. A corollary of this speculation is that if we ever detect another operational, active intelligence in the galaxy besides our own, it doubles the estimate of the lifetime of an advanced civilization, bringing it in the ballpark of 200,000 years.

This is simply a thought exercise, depending on very "unknown unknowns"—such as the probability of primitive life arising if the basic ingredients are present—and one can interpret it in a number of ways. The most obvious conclusion, for me, is that an advanced species does not necessarily receive an unlimited lease on life. If the cosmos is so silent, the odds must be stacked against it; in other words, in order to survive, a civilization has to work hard—and be somewhat lucky—to merit it. Preserving the human species and the biodiversity that supports it is not only a desirable option, from an egocentric perspective, it is also a respectful attitude toward the 14 billion years of cosmic evolution that brought us to where we are. Let us not throw it all away.

Notes

Chapter One

1. See Stilianos Louca et al., "A Census-Based Estimate of Earth's Bacteria and Archaeal Diversity," *PLoS Biology* 17, no. 2 (2019): e3000106, https://doi .org/10.1371/journal.pbio.3000106.

2. For more on this tally, see Camillo Mora et al., "How Many Species Are There on Earth and in the Ocean?," *PLoS Biology* 9, no. 8 (2011). e1001127, https://doi.org/10.1371/journal.pbio.1001127.

3. Norman Newell, "Revolutions in the History of Life," *Uniformity and Simplicity*, Geology Society of America, 89 (1967): 63–91, https://doi.org/10 .1130/SPE89-p63.

4. David M. Raup and J. John Sepkoski Jr. summarized the findings in "Mass Extinctions in the Marine Fossil Record," *Science* 215, no. 4539 (1982): 1501–3, https://doi.org/10.1126/science.215.4539.1501.

5. "Trap" comes from the Swedish word *trapp* ("stairway"), because these great plateaus of lava, when dismantled by erosion, reveal their layering as great steps in the landscape. The name has stuck for many of these large igneous provinces, such as the Siberian traps and the Deccan Traps.

Chapter Two

1. Alvarez et al., "Extraterrestrial Cause for the Cretaceous-Tertiary Extinction," *Science* 208, no. 4448 (1980): 1095–1108, https://doi.org/10.1126 /science.208.4448.1095. This article is a great example of how to build a scientific theory—based on data collection and analysis, reviewing alternate explanations, and predicting what further work should be undertaken to confirm the theory.

2. For an article summarizing the volcano theory, see Vincent Courtillot

et al., "Deccan Flood Basalts at the Cretaceous/Tertiary Boundary?," *Earth and Planetary Science Letters* 80, nos. 3–4 (1986): 361–74, https://doi.org/10.1016/0012-821X(86)90118-4.

3. For the discovery of impact-shocked quartz grains and impact-molten spherules, see Bruce Bohor et al., "Mineralogic Evidence for an Impact Event at the Cretaceous-Tertiary Boundary," *Science* 224, no. 4651 (1984): 867–69, https://doi.org/10.1126/science.224.4651.867, and Jan Smit and Gerald Klaver, "Sanidine Spherules at the Cretaceous-Tertiary Boundary Indicate a Large Impact Event," *Nature* 292 (1981): 47–49, https://doi.org/10.1038/292047a0.

4. For the evidence of Deccan volcanism leading up to the K-Pg crisis, see Anne-Lise Chenet et al., "Determination of Rapid Deccan Eruptions across the Cretaceous-Tertiary Boundary Using Paleomagnetic Secular Variation: 2. Constraints from Analysis of Eight New Sections and Synthesis for a 3500-m-Thick Composite Section," *Journal of Geophysical Research: Solid Earth* 114, no. B6 (2009), https://doi.org/10.1029/2008JB005644; and Xue Gu et al., "Deccan Volcanic Activity and Its Links to the End-Cretaceous Extinction in Northern China," *Global and Planetary Change* 210 (2022): 103772, https://doi.org/10.1016/j.gloplacha.2022.103772.

5. Concerning the claim that Deccan volcanism slowed down the recovery of the biosphere, see Paul Renne et al., "State Shift in Deccan Volcanism at the Cretaceous-Paleogene Boundary, Possibly Induced by an Impact," *Science* 350, no. 6256 (2015): 76-78, https://doi.org/10.1126/science.aac7549; and Courtney Sprain et al., "The Eruptive Tempo of Deccan Volcanism in Relation to the Cretaceous-Paleogene Boundary," *Science* 363, no. 6429 (2019): 866–70. For a rebuttal as to the purported negative effect of Deccan volcanism on the biosphere, see Alfio Alessandro Chiarenza et al., "Asteroid Impact, Not Volcanism, Caused the End-Cretaceous Dinosaur Extinction," *PNAS* 117, no. 29 (2020): 17084–93, https://doi.org/10.1073/pnas.2006087117. For the view that a small extinction did occur 150,000 years prior to impact, see Thomas Tobin et al., "Extinction Patterns, δ^{18} O Trends, and Magnetostratigraphy from a Southern High-Latitude Cretaceous-Paleogene Section: Links with Deccan Volcanism," *Palaeogeography, Palaeoclimatology, Palaeoecology* 350–352 (2012): 180–88.

6. For an instructive account of the K-Pg research and crater's discovery, see Alan Hildebrand, "The Cretaceous/Tertiary Boundary Impact (or the Dinosaurs Didn't Have a Chance)," *Journal of the Royal Astronomical Society of Canada* 87, no. 2 (1993): 77–118 (consultable through https://adsabs.harvard.edu/full/1993jrasc..87 . . . 77h). See also Walter Alvarez, *T. Rex and the Crater of Doom* (Princeton, NJ: Princeton University Press, 1997).

7. Carlos Byars, "Mexican Site May Be Link to Dinosaurs' Disappearance," *Houston Chronicle*, December 13, 1981, 1, 18.

8. Alan R. Hildebrand et al., "Chicxulub Crater: A Possible Cretaceous-Tertiary Boundary Impact Crater on the Yucatán Peninsula, Mexico," *Ge-*

ology 19, no. 9 (1991): 867–71, https://doi.org/10.1130/0091-7613(1991)019
<0867:CCAPCT>2.3.CO;2.

9. For a description of Chicxulub crater, see Jo Morgan et al., "Size and
Morphology of the Chicxulub Impact Crater," *Nature* 390 (1997): 472–76,
https://doi.org/10.1038/37291.

10. At the time of the dating, in 1992, the figures quoted were on the order
of 65 million years. Calibration improvements in radiochronology techniques
have since shifted all figures back to 66 million years.

11. Carl C. Swisher III et al., "Coeval ^{40}Ar/^{39}Ar Ages of 65.0 Million Years
Ago from Chicxulub Crater Melt Rock and Cretaceous-Tertiary Bound-
ary Tektites," *Science* 257, no. 5072 (1992): 954–58, https://doi.org/10.1126
/science.257.5072.954.

12. This "petering out" of rare fossils, before their true extinction level, is a
statistical bias, known as the Signor-Lipps effect, named after paleontologists
Philip Signor and Jere Lipps, who stressed its importance. Because there is so
much vertical distance (i.e., time separation) between rare fossils in a column
of sediments, there is little chance of finding a fossil close to the species' exact
time of disappearance. Last occurrences often fall short of the species' true
extinction level to make extinctions across the board look progressive rather
than sudden.

13. These are ballpark figures. Officially, the temperature unit used in physics and in these simulations is the kelvin (K), but since 1 kelvin is the same in-
crement as 1°C, I used Celsius in the text, which is more familiar to the reader
(with Fahrenheit in parentheses).

14. For details on the tsunami simulation, see Molly M. Range et al., "The
Chicxulub Impact Produced a Powerful Global Tsunami," *AGU Advances* 3,
no. 5 (2022): e2021AV000627, https://doi.org/10.1029/2021AV000627.

15. See Jan Smit et al., "Tektite-Bearing, Deep-Water Clastic Unit
at the Cretaceous-Tertiary Boundary in Northeastern Mexico," *Geology* 20, no. 2 (1992): 99–103, https://doi.org/10.1130/0091-7613(1992)020
<0099:TBDWCU>2.3.CO;2.

16. For a detailed description of the Tanis site, see Robert A. DePalma
et al., "A Seismically Induced Onshore Surge Deposit at the KPg Boundary,
North Dakota," *PNAS* 116, no. 17 (2019): 8190–99, https://doi.org/10.1073
/pnas.1817407116. For the evidence that the impact occurred in the spring, see
Melanie A. D. During et al., "The Mesozoic Terminated in Boreal Spring," *Nature* 603 (2022): 91–94, https://doi.org/10.1038/s41586-022-04446-1.

17. For a detailed look at the heat pulse of the Chicxulub impact, see Ta-
mara J. Goldin and H. Jay Melosh, "Self-Shielding of Thermal Radiation by
Chicxulub Impact Ejecta: Firestorm or Fizzle?," *Geology* 37, no. 12 (2009):
1135–38, https://doi.org/10.1130/G30433A.1.

18. The initial landmark article about plant combustion evidence at
the K-Pg boundary comes from Wendy Wolbach et al., "Global Fire at the

Cretaceous-Tertiary Boundary," *Nature* 334 (1988): 665–69, https://doi.org /10.1038/334665a0. For a later discussion, see Douglas S. Robertson et al., "K-Pg Extinction: Reevaluation of the Heat-Fire Hypothesis," *Journal of Geophysical Research: Biogeosciences* 118, no. 1 (2013): 329–36.

19. For an initial description of the K-Pg fern spike, see Robert H. Tschudy et al., "Disruption of the Terrestrial Plant Ecosystem at the Cretaceous-Tertiary Boundary, Western Interior," *Science* 225, no. 4666 (1984): 1030–32, https://doi.org/10.1126/science.225.4666.1030.

20. Charles G. Bardeen et al., "On Transient Climate Change at the Cretaceous–Paleogene Boundary Due to Atmospheric Soot Injections," *PNAS* 114, no. 36 (2017): E7415–E7424, https://doi.org/10.1073/pnas.1708980114.

21. See Johan Vellekoop et al., "Rapid Short-Term Cooling following the Chicxulub Impact at the Cretaceous-Paleogene Boundary," *PNAS* 111, no. 21 (2014): 7537–41, https://doi.org/10.1073/pnas.1319253111.

Chapter Three

1. For a description of the sterile "Strangelove ocean" in the wake of the impact, see Kenneth J. Hsu et al., "Mass Mortality and Its Environmental and Evolutionary Consequences," *Science* 216, no. 4543 (1982): 249–56, https:// doi.org/10.1126/science.216.4543.249. For a more detailed assessment of the marine kill, see Steven D'Hondt, "Consequences of the Cretaceous/ Paleogene Mass Extinction for Marine Ecosystems," *Annu. Rev. Ecol. Evol. Syst.* 36, no. 1 (2005): 295–317, https://doi.org/10.1146/annurev.ecolsys.35 .021103.105715.

2. The unit "pH," which stands for "power of hydrogen," is a scale to measure the acidity or basicity of water (or any aqueous solution). Solutions with a pH less than 7 are said to be acidic, and those with a pH over 7 are said to be basic (or alkaline). The scale is logarithmic, so that one pH unit corresponds to a tenfold difference in hydrogen ion concentration (i.e., in acidity or basicity).

3. For research done into the ocean's pH before and after the K-Pg impact, see Michael J. Henehan et al., "Rapid Ocean Acidification and Protracted Earth System Recovery Followed the End-Cretaceous Chicxulub Impact," *PNAS* 116, no. 45 (2019): 22500–22504, https://doi.org/10.1073/pnas .1905989116.

4. For the Colorado Springs record of recovery, see Tyler Lyson et al., "Exceptional Continental Record of Biotic Recovery after the Cretaceous-Paleogene Mass Extinction," *Science* 366, no. 6438 (2019): 977–83, https://doi .org/10.1126/science.aay2268.

5. For more on *Purgatorius*, see Gregory P. Wilson Mantilla et al., "Earliest Palaeocene Purgatoriids and the Initial Radiation of Stem Primates," *R. Soc. Open Sci.* 8, no. 2 (2021): 210050, https://doi.org/10.1098/rsos.210050.

6. See James Kennett and Lowell Stott, "Abrupt Deep-Sea Warming, Pa-

laeoceanographic Changes and Benthic Extinctions at the End of the Palaeo-cene," *Nature* 353 (1991): 225–29, https://doi.org/10.1038/353225a0.

7. For more details on the methane release of sea-bottom sediment as a trigger of the PETM warm period, see Henrick Svensen et al., "Release of Methane from a Volcanic Basin as a Mechanism for Initial Eocene Global Warming," *Nature* 429 (2004): 542–45, https://doi.org/10.1038/nature02566. Concerning the volcanic triggering of the PETM warm period, see Thomas M. Gernon et al., "Transient Mobilization of Subcrustal Carbon Coincident with Palaeocene–Eocene Thermal Maximum," *Nat. Geosci.* 15 (2022): 573–79, https://doi.org/10.1038/s41561-022-00967-6; and Tali Babila et al., "Surface Ocean Warming and Acidification Driven by Rapid Carbon Release Precedes Paleocene-Eocene Thermal Maximum," *Sci. Adv.* 8, no. 11 (2022): eabg1025, https://doi.org/10.1126/sciadv.abg1025.

8. For a review of warming events such as the PETM and global warming today, see Gavin L. Foster et al., "Placing Our Current 'Hyperthermal' in the Context of Rapid Climate Change in Our Geological Past," *Philos. Trans. R. Soc. A: Math. Phys. Eng. Sci.* 376, no. 2130 (2018): 20170086, http://dx.doi.org/10.1098/rsta.2017.0086.

9. See Francesca A. McInerney and Scott L. Wing, "The Paleocene-Eocene Thermal Maximum: A Perturbation of Carbon Cycle, Climate, and Biosphere with Implications for the Future," *Annu. Rev. Earth & Planet. Sci.* 39 (2011): 489–516, https://doi.org/10.1146/annurev-earth-040610-133431.

10. To my chagrin, I was never able to find or remember the source of that article, which I read in 1978.

11. Charles Frankel, *Quatre milliards d'années d'histoire de la Terre* (Paris: Ed. De Vecchi, 1980).

12. See René Bobe, Anna K. Behrensmeyer, and Ralph E. Chapman, "Faunal Change, Environmental Variability and Late Pliocene Hominin Evolution," *Journal of Human Evolution* 42, no. 4 (2002): 475–97, https://doi.org/10.1006/jhev.2001.0535.

13. See Brian Villmore et al., "Early *Homo* at 2.8 Ma from Ledi-Geraru, Afar, Ethiopia," *Science* 347, no. 6228 (2015): 1352–55, https://doi.org/10.1126/science.aaa1343.

14. Continents are shifted around the Earth's surface by plate tectonics, but their distribution can also constitute a mass imbalance on the rotating Earth and go so far as to shift its rotation axis: continents are shifted with respect to the poles, a phenomenon known as *true polar wander*.

Chapter Four

1. For the landmark article charging humankind with the slaughter of the megafauna, and a more recent assessment by the same author, see Paul Martin, "Prehistoric Overkill," in *Pleistocene Extinctions: The Search for a Cause,*

ed. P. S. Martin and H. E. Wright Jr. (New Haven, CT: Yale University Press, 1967), 75–120; and Paul S. Martin, "40,000 Years of Extinctions on the 'Planet of Doom,'" *Palaeogeography, Palaeoclimatology, Palaeoecology* 82, nos. 1–2 (1990): 187–201, https://doi.org/10.1016/S0031-0182(12)80032-0.

2. For a view of early extinctions of the megafauna in Africa, see J. Tyler Faith, "Late Pleistocene and Holocene Mammal Extinctions on Continental Africa," *Earth-Science Reviews* 128 (2014): 105–21, https://doi.org/10.1016/j .earscirev.2013.10.009; and J. Tyler Faith et al., "Plio-Pleistocene Decline of African Megaherbivores: No Evidence for Ancient Hominine Impacts," *Science* 362, no. 6417 (2018): 938–41, https://doi.org/10.1016/10.1126/science.aau2728.

3. See Francis Wenban-Smith et al., "The Clactonian Elephant Butchery Site at Southfleet Road, Ebbsfleet, UK," *Journal of Quaternary Science* 21, no. 5 (2006): 471–83, https://doi.org/10.1002/jqs.1033; and Hartmut Thieme and Stephan Veil, "Neue Untersuchungen zum eemzeitlichen Elefanten-Jagdplatz Lehringen, Ldkr. Verden," *Kunde* 36 (1985): 11–58.

4. For a discussion on the disappearance of the giant elk, see Adrian Lister and Anthony Stuart, "The Extinction of the Giant Deer *Megaloceros giganteus* (Blumenbach): New Radiocarbon Evidence," *Quaternary evidence* 500 (2019): 185–203.

5. See Richard Firestone et al., "Evidence for an Extraterrestrial Impact 12,900 Years Ago That Contributed to the Megafaunal Extinctions and the Younger Dryas Cooling," *PNAS* 104, no. 41 (2007): 16016–21.

6. See Wendy Wolbach et al., "Extraordinary Biomass-Burning Episode and Impact Winter Triggered by the Younger Dryas Cosmic Impact ~12,800 Years Ago. 1. Ice Cores and Glaciers," *Journal of Geology* 126, no. 2 (2018): 165–84, https://doi.org/10.1086/695703.

7. Martin Sweatman, "The Younger Dryas Impact Hypothesis: Review of the Impact Evidence," *Earth-Science Reviews* 218 (2021): 103677, https://doi .org/10.1016/j.earscirev.2021.103677.

8. For the "end of the world" Clovis site, see Guadalupe Sanchez et al., "Human (Clovis)-Gomphothere (*Cuvieronius* sp.) Association ~13,390 Calibrated yBP in Sonora, Mexico," *PNAS* 111, no. 30 (2014): 10972–77, https:// doi.org/10.1073/pnas.1404546111.

9. See Gary Haynes, "A Review of Some Attacks on the Overkill Hypothesis, with Special Attention to Misrepresentation and Doubletalk," *Quaternary International* 169–170 (2006): 84–94, https://doi.org/10.1016/j.quaint.2006 .07.002.

10. See David Bustos et al., "Footprints Preserve Terminal Pleistocene Hunt? Human-Sloth Interactions in North America," *Science Advances* 4, no. 4 (2018): eaar7621, https://doi.org/10.1126/sciadv.aar7621.

11. See Jacquelyn Gill et al., "Pleistocene Megafaunal Collapse, Novel Plant Communities, and Enhanced Fire Regimes in North America," *Science* 326, no. 5956 (2009): 1100–1103, https://doi.org/10.1126/science.1179504.

12. See Susan Rule et al., "The Aftermath of Megafaunal Extinction: Eco-system Transformation in Pleistocene Australia," *Science* 335, no. 6075 (2012): 1483–86, https://doi.org/10.1126/science.1214261; and Christopher Johnson et al., "Using Dung Fungi to Interpret Decline and Extinction of Megaherbivo-res: Problems and Solutions," *Quaternary Science Reviews* 110 (2015): 107–13, https://doi.org/10.1016/j.quascirev.2014.12.011.

13. See Christopher Johnson, "Determinants of Loss of Mammal Species during the Late Quaternary 'Megafauna' Extinctions: Life History and Ecol-ogy, but Not Body Size," *Proc. R. Soc. Lond. B* 269, no. 1506 (2002): 2221–27, https://doi.org/10.1098/rspb.2002.2130; and Christopher Johnson, "Megafau-nal Decline and Fall," *Science* 326, no. 5956 (2009): 1072–73, https://doi.org/10.1126/science.1182770.

14. See Ross McPhee, M. A. Iturralde-Vinent, and Osvaldo Jiménez Vázquez, "Prehistoric Sloth Extinctions in Cuba: Implications of a New 'Last' Appearance Date," *Caribbean Journal of Science* 43, no. 1 (2007): 94–98, https://doi.org/10.18475/cjos.v43i1.a9.

15. See Allison Karp et al., "Global Response of Fire Activity to Late Qua-ternary Grazer Extinctions," *Science* 374, no. 6571 (2021): 1145–48, https://doi.org/10.1126/science.abj1580.

Chapter Five

1. One pigmy sloth survived extinction on a small island off the coast of Panama, Isla Escudo de Veraguas, where it is now protected. For a review of the Caribbean giant sloth extinctions, see David Steadman et al., "Asynchro-nous Extinction of Late Quaternary Sloths on Continents and Islands," *PNAS* 102, no. 33 (2005): 11763–68, https://doi.org/10.1073/pnas.0502777102.

2. See David Burney et al., "*Sporormiella* and the Late Holocene Extinc-tions in Madagascar," *PNAS* 100, no. 19 (2003): 10800–10805, https://doi.org/10.1073/pnas.1534700100.

3. Blaming the demise of the moa birds in New Zealand on a cometary impact relied on a 20-kilometer (12 mi.) crater on the seafloor, off the coast of South Island, and on tsunami deposits. Both the impact origin for the forma-tion and the age of the tsunami deposits have been challenged.

4. For the extinction history of New Zealand's moa birds, see Morten Al-lentoft et al., "Extinct New Zealand Megafauna Were Not in Decline before Human Colonization," *PNAS* 111, no. 13 (2014): 4922–27, https://doi.org/10.1073/pnas.1314972111; and George Perry et al., "A High-Precision Chronol-ogy for the Rapid Extinction of New Zealand Moa (Aves, Dinornithiformes)," *Quaternary Science Reviews* 105 (2014): 126–35, https://doi.org/10.1016/j.quascirev.2014.09.025.

5. See Aaron Bauer and Anthony Russell, "*Hoplodactylus delcourti* n. sp. (Reptilia: Gekkonidae), the Largest Known Gecko," *New Zealand Journal*

of Zoology 13, no. 1 (1986): 141–48, https://doi.org/10.1080/03014223.1986 .10422655.

6. The passenger pigeon is distinct from the carrier pigeon (*Columba livia*) and is characterized by a smaller head, orange to bronze-colored feathers on the neck (for the male), and narrower, longer wings.

7. Paul Ehrlich, David S. Dobkin, and Darryl Wheye, *The Birder's Handbook* (New York: Simon and Schuster, 1988), 277.

8. For a tally of bird species (and species in general), see the IUCN list, https://www.iucnredlist.org/statistics. Also see the website of the International Ornithologist Congress (IOC), https://www.worldbirdnames.org/, along with its World Bird List (revised every six months): Frank Gill, David Donsker, and Pamela Rasmussen, eds., IOC World Bird List (v. 13.1), 2023, https://doi.org/10.14344/IOC.ML.13.1.

9. Before modern times, the extinction of species was more or less balanced by the appearance of new ones, which branched off from each other. This creative rate of divergence is naturally slow and cannot counterbalance the much faster-paced losses inflicted by humans.

10. See Richard Duncan, Alison Boyer, and Tim Blackburn, "Magnitude and Variation of Prehistoric Bird Extinctions in the Pacific," *PNAS* 110, no. 16 (2013): 6436–41, https://doi.org/10.1073/pnas.1216511110.

11. For a tally of mammals, see the website of the American Society of Mammalogists, https://www.mammaldiversity.org, and the study by Connor J. Burgin et al., "How Many Species of Mammals Are There?" *Journal of Mammalogy* 99, no. 1 (2018): 1–14, https://doi.org/10.1093/jmammal/gyx147.

12. Lists of new species and other environmental and wildlife news can be found on Mongabay's website: https://news.mongabay.com/list/new-species/.

13. For the extinction of the baiji dolphin, see Samuel Turvey et al., "First Human-Caused Extinction of a Cetacean Species?," *Biology Letters* 3 (2007): 537–40, https://doi.org/10.1098/rsbl.2007.0292.

14. For a current tally of reptiles, see the Reptile Database, http://www .reptile-database.org/.

15. See Corentin Bochaton et al., "Large-Scale Reptile Extinctions following European Colonization of the Guadeloupe Islands," *Science Advances* 7, no. 21 (2021): eabg2111, https://doi.org/10.1126/sciadv.abg2111.

16. For a current tally of amphibians, see AmphibiaWeb, http://www .amphibiaweb.org/.

17. For a tally of recently extinct amphibians, as well as other statistics for all animal, fungi, and plant species, see the IUCN's website, https://www .iucnredlist.org/statistics.

18. For the tally of mollusks, see Robert Cowie et al., "Measuring the Sixth Extinction: What Do Mollusks Tell Us?," *The Nautilus* 131, no. 1 (2017): 3–41.

19. For the tally of fish, see "Summary Statistics: Table 3," IUCN, https:// www.iucnredlist.org/statistics.

20. For the extinction of the smooth handfish (*Sympterichthys unipennis*), see Elizabeth Claire Alberts, "The First Modern-Day Marine Fish Has Officially Gone Extinct. More May Follow," Mongabay, June 26, 2020, https://news.mongabay.com/2020/06/the-first-modern-day-marine-fish-has-officially-gone-extinct-more-may-follow/.

21. See Cowie et al., "Measuring the Sixth Extinction."

22. "Greatest Concentration of Insects," Guinness World Records, accessed February 21, 2023, https://www.guinnessworldrecords.com/world-records/70681-greatest-concentration-of-insects.

23. For the full story of the Rocky Mountain locust and the causes of its extinction, see Jeffrey Lockwood, *Locust: The Devastating Rise and Mysterious Disappearance of the Insect That Shaped the American Frontier* (New York: Basic Books, 2004).

24. See Matthew Trager and Jaret Daniels, "Ant Tending of Miami Blue Butterfly Larvae (Lepidoptera: Lycaenidae): Partner Diversity and Effects on Larval Performance," *Florida Entomologist* 92, no. 3 (2009): 474–82, https://doi.org/10.1653/024.092.0309.

25. See David Hawksworth and Robert Lücking, "Fungal Diversity Revisited: 2.2 to 3.8 Million Species," *Microbiology Spectrum* 5, no. 4 (2017), https://doi.org/10.1128/microbiolspec.FUNK-0052-2016; and Martin Cheek et al., "New Scientific Discoveries: Plants and Fungi," *Plants, People, Planet* 2, no. 5 (2020): 371–88, https://doi.org/10.1002/ppp3.10148.

26. See Gregory Mueller et al., "What Do the First 597 Global Fungal Red List Assessments Tell Us about the Threat Status of Fungi," *Diversity* 14, no. 9 (2022): 736, https://doi.org/10.3390/d14090736.

27. For an assessment of plant diversity, see "State of the World's Plants and Fungi," Royal Botanic Gardens, Kew, accessed February 22, 2023, https://www.kew.org/science/state-of-the-worlds-plants-and-fungi; and Martin Cheek et al., "New Scientific Discoveries: Plants and Fungi," *Plants, People, Planet* 2, no. 5 (2020): 371–88, https://doi.org/10.1002/ppp3.10148.

28. For a status of plant and fungi threats and extinctions, see Eimear Nic Lughadha et al., "Extinction Risk and Threats to Plants and Fungi," *Plants, People, Planet* 2, no. 5 (2020): 389–408, https://doi.org/10.1002/ppp3.10146.

29. See Quentin Cronk, "Plant Extinctions Take Time," *Science* 353, no. 6298 (2016): 446–47, https://doi.org/10.1126/science.aag1794.

Chapter Six

1. Rodolfo Dirzo et al., "Defaunation in the Anthropocene," *Science* 345, no. 6195 (2014): 401–406, https://doi.org/10.1126/science.1251817.

2. WWF, *Living Planet Report 2022—Building a Nature-Positive Society*, ed. Almond et al. (Gland, Switzerland: WWF, 2022), https://wwflpr.awsassets.panda.org/downloads/lpr_2022_full_report.pdf.

3. Michael Chase et al., "Continent-Wide Survey Reveals Massive Decline in African Savannah Elephants," *PeerJ* 4 (2016): e2354, https://doi.org/10.7717/peerj.2354.

4. See Barb Dean-Simmons, "Cod Stocks on Newfoundland's South Coast Are Growing, but Still in 'Critical Zone,'" Saltwire, November 19, 2021, https://www.saltwire.com/atlantic-canada/business/cod-stocks-on-newfoundlands-south-coast-are-growing-but-still-in-critical-zone-100660449/.

5. For a tally of Amazon species, see Rhett A. Butler, "Animals of the Amazon Rainforest," Mongabay, April 1, 2019, https://rainforests.mongabay.com/amazon/amazon_wildlife.html.

6. For more on the concept of "extinction debt," see David Tilman et al., "Habitat Destruction and the Extinction Debt," *Nature* 371 (1994): 65–66, https://doi.org/10.1038/371065a0; for its application to the Amazon rainforest, see Oliver Wearn et al., "Extinction Debt and Windows of Conservation Opportunity in the Brazilian Amazon," *Science* 337, no. 6091 (2012): 228–32, https://doi.org/10.1126/science.1219013.

7. "Extinction of Rainforest Species Slows Future Growth," Rainforest-Maker blog, accessed February 23, 2023, https://www.rainforestmaker.org/extinction-of-rainforest-species-slows-future-growth.html.

8. The IUCN's Red List of Threatened Species can be found at https://www.iucnredlist.org. See its summary statistics and tables: https://www.iucnredlist.org/resources/summary-statistics.

9. IUCN, *The IUCN Red List of Threatened Species*, version 2022-2, accessed August 15, 2022, https://www.iucnredlist.org.

10. See Su Shiung Lam et al., "Environmental Management of Two of the World's Most Endangered Marine and Terrestrial Predators: Vaquita and Cheetah," Environmental Research 190 (2020): 109966, https://doi.org/10.1016/j.envres.2020.109966; and Gabrielle Canon, "There Are Fewer than 10 Tiny Vaquita Porpoises Left. Can They Be Saved?," *The Guardian*, February 12, 2022, https://www.theguardian.com/world/2022/feb/11/tiny-vaquita-numbers-less-than-10-can-they-be-saved.

11. All figures are derived from IUCN Red List tables and summaries, found at https://www.iucnredlist.org/resources/summary-statistics (accessed August 15, 2022); any errors in computing totals and percentages are mine alone.

12. See Anthony Barnosky et al., "Has the Earth's Sixth Mass Extinction Already Arrived?," *Nature* 471 (2011): 51–57, https://doi.org/10.1038/nature09678.

Chapter Seven

1. The control of the Earth's climate by lifeforms was presented by British scientist James Lovelock (1919–2022) as a planetary-scale homeostasis model,

known as the Gaia hypothesis (named after the Greek goddess personifying the Earth).

2. See Felisa A. Smith, Scott A. Elliott, and S. Kathleen Lyons, "Methane Emissions from Extinct Megafauna," *Nature Geoscience* 3 (2010): 374–75, https://doi.org/10.1038/ngeo877; and Felisa Smith et al., "Exploring the Influence of Ancient and Historic Megaherbivore Extirpations on the Global Methane Budget," PNAS 113, no. 4 (2016): 874–79, https://doi.org/10.1073/pnas.1502547112.

3. IPCC, *Climate Change 1995: The Science of Climate Change*, ed. J. T. Houghton et al. (Cambridge: Cambridge University Press, 1996), https://www.ipcc.ch/site/assets/uploads/2018/02/ipcc_sar_wg_I_full_report.pdf.

4. IPCC, *Climate Change 1995*.

5. IPCC, *Climate Change 2001: Synthesis Report*, ed. Robert T. Watson et al. (Cambridge: Cambridge University Press, 2001), https://www.ipcc.ch/site/assets/uploads/2018/05/SYR_TAR_full_report.pdf.

6. IPCC, *Climate Change 2007: Impacts, Adaptation and Vulnerability*, ed. Martin Parry et al. (Cambridge: Cambridge University Press, 2007), https://www.ipcc.ch/site/assets/uploads/2018/03/ar4_wg2_full_report.pdf.

7. IPCC, *Climate Change 2013: The Physical Science Basis*, ed. Thomas F. Stocker et al. (Cambridge: Cambridge University Press, 2013), https://www.ipcc.ch/site/assets/uploads/2018/03/WG1AR5_SummaryVolume_FINAL.pdf.

8. IPCC, *Climate Change 2021: The Physical Science Basis*, ed. Valérie Masson-Delmotte et al. (Cambridge: Cambridge University Press, 2021), https://www.ipcc.ch/report/ar6/wg1/downloads/report/IPCC_AR6_WGI_FullReport_small.pdf.

9. See Alan Pounds and Martha Crump, "Amphibian Declines and Climate Disturbance: The Case of the Golden Toad and the Harlequin Frog," *Conservation Biology* 8, no. 1 (1994): 72–85, https://doi.org/10.1046/j.1523-1739.1994.08010072.x; and Alan Pounds et al., "Biological Response to Climate Change on a Tropical Mountain," *Nature* 398 (1999): 611–15, https://doi.org/10.1038/19297.

10. See Richard Kock et al., "Saigas on the Brink: Multidisciplinary Analysis of the Factors Influencing Mass Mortality Events," *Science Advances* 4, no. 1 (2018), https://doi.org/10.1126/sciadv.aao2314.

11. See Brian Beckage et al., "A Rapid Upward Shift of a Forest Ecotone during 40 Years of Warming in the Green Mountains of Vermont," *PNAS* 105, no. 11 (2008): 4197–202, https://doi.org/10.1073/pnas.0708921105.

12. Joseph Grinnell and Tracy Irwin Storer, *Animal Life in the Yosemite: An Account of the Mammals, Birds, Reptiles, and Amphibians in a Cross-Section of the Sierra Nevada* (Berkeley: University of California Press, 1924). Joseph Grinnell (1877–1939) was one of the first zoologists to study ecological communities and developed the concept of ecological niches.

13. Craig Moritz et al., "Impact of a Century of Climate Change on Small-Mammal Communities in Yosemite Natural Park, USA," *Science* 322, no. 5899 (2008): 261–64, https://doi.org/10.1126/science.1163428.

14. See Camille Parmesan and Gary Yohe, "A Globally Coherent Fingerprint of Climate Change Impacts across Natural Systems," *Nature* 421, no. 6918 (2003): 37–42, https://doi.org/10.1038/nature01286.

15. There is a major difference, however. During transitions out of glacial stages, temperatures rise from values *lower* than average, as they gain 4°C to 5°C (7.2°F to 9°F). Today, we are already in the warm part of the cycle—an interglacial stage—so adding an extra 2°C to 4°C (3.6°F to 7.2°F), in an intermediate global warming scenario, brings the temperature above values experienced over the past 50 million years.

16. See Scott Loarie et al., "The Velocity of Climate Change," *Nature* 462, no. 7276 (2009): 1052–55, https://doi.org/10.1038/nature08649.

17. See Ghislain Vieilledent et al., "Vulnerability of Baobab Species to Climate Change and Effectiveness of the Protected Area Network in Madagascar: Towards New Conservation Priorities," *Biological Conservation* 166 (2013): 11–22, https://doi.org/10.1016/j.biocon.2013.06.007.

18. See Leydemire Oliveira et al., "Large-Scale Expansion of Agriculture in Amazonia May Be a No-Win Scenario," *Environ. Res. Lett.* 8, no. 2 (2013): 024021, https://doi.org/10.1088/1748-9326/8/2/024021.

19. See Paulo Brando et al., "Abrupt Increases in Amazonian Tree Mortality Due to Drought-Fire Interactions," *PNAS* 111, no. 17 (2014): 6347–52, https://doi.org/10.1073/pnas.1305499111.

20. For more on sea-level rise projections, see "2022 Sea Level Rise Technical Report," National Oceanic and Atmospheric Administration, accessed February 27, 2023, https://oceanservice.noaa.gov/hazards/sealevelrise/sea levelrise-tech-report.html.

21. See "Coral Reefs 101: Reef Threats," Coral Reef Alliance, accessed February 27, 2023, https://coral.org/en/coral-reefs-101/reef-threats/; "Coral Reefs," Endangered Species International, accessed February 27, 2023, https://www.endangeredspeciesinternational.org/coralreefs4.html; and Kent Carpenter et al., "One-Third of Reef-Building Corals Face Elevated Extinction Risk from Climate Change and Local Impacts," *Science* 321, no. 5888 (2008): 560–63, https://doi.org/10.1126/science.1159196.

22. See "Ocean Acidification," European Environment Agency, April 28, 2023, https://www.eea.europa.eu/ims/ocean-acidification; and "Ocean Acidification," Smithsonian, April 2018, https://ocean.si.edu/ocean-life /invertebrates/ocean-acidification.

23. See Glenn De'ath, Janice M. Lough, and Katharina E. Fabricius, "Declining Coral Calcification on the Great Barrier Reef," *Science* 323, no. 5910 (2009): 116–19, https://doi.org/10.1126/science.1165283.

24. See K. J. Sebastian Meier et al., "The Role of Ocean Acidification in

Emiliania huxleyi Coccolith Thinning in the Mediterranean Sea," *Biogeosciences* 11, no. 10 (2014): 2857–69, https://doi.org/10.5194/bg-11-2857-2014.

25. See Henry Holm et al.: "Global Ocean Lipidomes Show a Universal Relationship between Temperature and Lipid Unsaturation," *Science* 376, no. 6600 (2022): 1487–91, https://doi.org/10.1126/science.abn7455.

Chapter Eight

1. See Julia Marton-Lefèvre, "Biodiversity Is Our Life," *Science* 327, no. 5970 (2010): 1179, https://doi.org/10.1126/science.1188424.

2. "Climate Change Killing Elephants, Says Kenya," BBC, July 28, 2022, video, 2:17, https://www.bbc.com/news/av/world-africa-62323058.

3. See "2021 Numbers (Unofficial) Indicate that Poaching/Trafficking Still Active in Kenya," SEEJ–AFRICA, March 21, 2022, https://www.seej-africa .org/kenya/2021-numbers-unofficial-indicate-that-poaching-trafficking-still -active-in-kenya/; and Jennifer Hassan, "Climate Change Is Killing More Elephants than Poaching, Kenyan Officials Say," *Washington Post*, July 28, 2022, https://www.washingtonpost.com/world/2022/07/28/kenya-elephants -drought-climate-change/.

4. See Shane Campbell-Staton et al., "Ivory Poaching and the Rapid Evolution of Tusklessness in African Elephants," *Science* 374, no. 6566 (2021): 483–87, https://doi.org/10.1126/science.abe7389.

5. Michael Hoffman et al., "The Impact of Conservation on the Status of the World's Vertebrates," *Science* 330, no. 6010 (2010): 1503–9, https://doi.org /10.1126/science.1194442.

6. IUCN Green Status of Species, https://www.iucnredlist.org/about /green-status-species.

7. See OECD, *Biodiversity: Finance and the Economic and Business Case for Action: A Report Prepared for the G7 Environment Ministers' Meeting, 5–6 May 2019* (OECD, 2019), https://www.oecd.org/environment/resources /biodiversity/G7-report-Biodiversity-Finance-and-the-Economic-and -Business-Case-for-Action.pdf; and The Bonn Challenge, https://www.bonn challenge.org.

8. See Shaden Khalifa et al., "Overview of Bee Pollination and Its Economic Value for Crop Production," *Insects* 12, no. 8 (2021): 688, https://doi .org/10.3390/insects12080688.

9. See Justin Boyles et al., "Economic Importance of Bats in Agriculture," *Science* 332, no. 6025 (2011): 41–42, https://doi.org/10.1126/science.1201366.

10. See Sara Kross et al., "Effects of Introducing Threatened Falcons into Vineyards on Abundance of Passeriformes and Bird Damage to Grapes," *Conservation Biology* 26, no. 1 (2012): 142–49, https://doi.org/10.1111/j.1523-1739 .2011.01756.x.

11. See Theodore Evans et al., "Ants and Termites Increase Crop Yield in a

Dry Climate," *Nature Communications* 2, no. 1 (2011): 262, https://doi.org/10 .1038/ncomms1257.

12. See Tyler Eddy et al., "Global Decline in Capacity of Coral Reefs to Provide Ecosystem Services," *One Earth* 4, no. 9 (2021): 1278–85, https://doi .org/10.1016/j.oneear.2021.08.016.

13. See Charles M. Peters, Alwyn H. Gentry, and Robert O. Mendelsohn, "Valuation of an Amazonian Rainforest," *Nature* 339, no. 6227 (1989): 655–56, https://doi.org/10.1038/339655a0.

14. See Guillaume Chapron et al., "Recovery of Large Carnivores in Europe's Modern Human-Dominated Landscapes," *Science* 346, no. 6216 (2014): 1517–19, https://doi.org/10.1126/science.1257553.

15. See the website of Colossal Biosciences, https://colossal.com /thylacine/; and Kate Evans, "De-extinction Company Aims to Resurrect the Tasmanian Tiger," *Scientific American*, August 16, 2022, https://www .scientificamerican.com/article/de-extinction-company-aims-to-resurrect -the-tasmanian-tiger/.

16. See Emma Dorey, "Huia Cloned Back to Life?," *Nature Biotechnology* 17 (1999): 736, https://doi.org/10.1038/11628.

Chapter Nine

1. See Ann Gibbons, "Pleistocene Population Explosions," *Science* 262, no. 5130 (1993): 27–28, https://doi.org/10.1126/science.262.5130.27; Michael Rampino and Stephen Self, "Climate-Volcanism Feedback and the Toba Eruption of ~74,000 Years Ago," *Quaternary Research* 40, no. 3 (1993): 269–80, https://doi.org/10.1006/qres.1993.1081; Stanley Ambrose, "Late Pleistocene Human Population Bottlenecks, Volcanic Winter, and Differentiation of Modern Humans," *Journal of Human Evolution* 34, no. 6 (1998): 623–51, https:// doi.org/10.1006/jhev.1998.0219; and Ann Gibbons, "Human Ancestors Were an Endangered Species," *Science News*, January 19, 2010, https://www.science .org/content/article/human-ancestors-were-endangered-species.

2. See Chad Yost et al., "Subdecadal Phytolith and Charcoal Records from Lake Malawi, East Africa Imply Minimal Effects on Human Evolution from the ~74 ka Toba Supereruption," *Journal of Human Evolution* 116 (2018): 75–94, https://doi.org/10.1016/j.jhevol.2017.11.005.

3. These figures refer to *effective population size*, a theoretical and simplified number of breeding individuals that fit the DNA data. The real population at the time might have been somewhat larger.

4. The hypothesis of a pandemic to explain the demise of the dinosaurs was put forward by American paleontologist Robert Bakker in the 1980s. See Robert Bakker, *The Dinosaur Heresies: New Theories Unlocking the Mystery of the Dinosaurs and Their Extinction* (New York: William Morrow, 1986).

5. The name "Spanish flu" comes from the fact that the epidemic was only

reported by the Spanish press at the time. Other European countries were at war (World War I) and censored the information in order not to further sap the morale of their troops.

6. Michael Worobey et al., "The Huanan Seafood Wholesale Market in Wuhan Was the Early Epicenter of the COVID-19 Pandemic," *Science* 377, no. 6609 (2022): 951–59, https://doi.org/10.1126/science.abp8715.

7. See Edward Holmes et al., "The Origins of SARS-CoV-2: A Critical Review," *Cell* 184, no. 19 (2021): 4848–56, https://doi.org/10.1016/j.cell.2021.08.017.

8. Although the intentional spread of viruses through objects like blankets was unlikely to work, the unintentional spread of diseases by European soldiers and settlers through close contact and respiratory transmission killed an estimated 90 percent of the pre-Columbian native population in the Americas. See also "1763–64: Britain Wages Biological Warfare with Smallpox," Native Voices, National Library of Medicine, accessed March 1, 2023, https://www.nlm.nih.gov/nativevoices/timeline/229.html.

9. The world total stockpile is closer to 13,000 nuclear bombs, if "retired" warheads are included.

10. For a perspective on these statistics, see Amy Nelson and Alexander Montgomery, "How Not to Estimate the Likelihood of Nuclear War," Brookings Institution, October 19, 2022, https://www.brookings.edu/blog/order-from-chaos/2022/10/19/how-not-to-estimate-the-likelihood-of-nuclear-war/.

11. The symbolic Doomsday Clock, established in 1947 and managed by the members of the *Bulletin of the Atomic Scientists*, estimates every year (on January 1) the threat to humanity (mostly due to atomic war and climate change) as a number of minutes or seconds leading up to midnight. A record was set on January 1, 2023, with the needle moved up to ninety seconds before midnight). See "Doomsday Clock," *Bulletin of the Atomic Scientists*, accessed March 1, 2023, https://thebulletin.org/doomsday-clock/.

12. See Owen B. Toon, Alan Robock, and Richard P. Turco, "Environmental Consequences of Nuclear War," *Physics Today* 61, no. 12 (2008): 37–42, https://doi.org/10.1063/1.3047679.

13. See Jonas Jägermeyr et al., "A Regional Nuclear Conflict Would Compromise Global Food Security," *PNAS* 117, no. 13 (2020): 7071–81, https://doi.org/10.1073/pnas.1919049117.

14. See Toon et al., "Environmental Consequences of Nuclear War"; and Joshua Coupe et al., "Nuclear Winter Responses to Nuclear War between the United States and Russia in the Whole Atmosphere Community Climate Model Version 4 and the Goddard Institute for Space Studies ModelE," *JGR Atmospheres* 124, no. 15 (2019): 8522–43, https://doi.org/10.1029/2019JD030509.

15. "Go Grandmaster Lee Se-Dol Retires Saying Artificial Intelligence

Cannot Be Defeated," ABC News, November 27, 2019, https://www.abc.net
.au/news/2019-11-28/go-grandmaster-lee-se-dol-retires-computers-cannot
-be-defeated/11745872.

16. Fabio Urbina et al., "Dual Use of Artificial-Intelligence-Powered Drug
Discovery," *Nature Machine Intelligence* 4 (2022): 189–91, https://doi.org/10
.1038/s42256-022-00465-9.

17. See Dennis Pamlin and Stuart Armstrong, *12 Risks That Threaten Human Civilisation: The Case for a New Risk Category*, report (Stockholm: Global
Challenges Foundation, 2015).

18. See NASA's DART web page, https://dart.jhuapl.edu, accessed March
1, 2023; and the press kit, "Double Asteroid Redirection Test," NASA, accessed March 1, 2013, https://dart.jhuapl.edu/News-and-Resources/files
/DART-press-kit-web-FINAL.pdf.

19. Elon Musk (@elonmusk), Twitter, February 10, 2019, 10:12 p.m.,
https://twitter.com/elonmusk/status/1094796246613516289?s=20.

20. Amanda Kooser, "Elon Musk Drops Details for SpaceX Mars Mega-
Colony," CNET, January 16, 2020, https://www.cnet.com/science/elon-musk
-drops-details-for-spacexs-million-person-mars-mega-colony/.

21. While a minimum population of 500 to 1,000 individuals seems suitable to ensure the genetic health of a Martian colony, founding the colony with
only two or three women is also possible, as they could use a sperm bank to
ensure the requisite genetic variability.

Index

Page numbers in italics refer to figures.